Signals and Communication Technology

More information about this series at http://www.springer.com/series/4748

Tudor Barbu

Novel Diffusion-Based Models for Image Restoration and Interpolation

 Springer

Tudor Barbu
Institute of Computer Science
  of the Romanian Academy
Iaşi, Romania

ISSN 1860-4862   ISSN 1860-4870   (electronic)
Signals and Communication Technology
ISBN 978-3-030-06566-9   ISBN 978-3-319-93006-0   (eBook)
https://doi.org/10.1007/978-3-319-93006-0

Printed on acid-free paper

This Springer imprint is published by the registered company Springer International Publishing AG
part of Springer Nature
The registered company address is: Gewerbestrasse 11, 6330 Cham, Switzerland

To my family

Contents

Chapter 1
Introduction

The mathematical models based on partial differential equations (PDEs) have been increasingly applied in some traditionally engineering domains, such as signal and image processing, image analysis and computer vision, for more than three decades. Since the partial differential equations express continuous change, they have long been used to formulate dynamical phenomena in some important engineering areas.

These equations have been successfully used for solving many image processing and computer vision tasks in these years, leading to an entire new subdomain of this scientific field. Thus, they offer some important advantages to image processing and analysis and computer vision domains, such as their modelling flexibility, their strong mathematical foundation, and their numerical approximations representing reinterpretations of numerous classical image processing techniques.

Many variational techniques for various image processing and computer vision have also been developed in the last 30 years, because it is very common in these fields to derive PDE-based models from some variational problems implying energy cost functional minimizations. The PDE variational approaches have some important advantages in both theory and computation, since they could achieve high speed, accuracy and stability also.

Besides the PDE schemes that follow variational principles, numerous PDE-based models which are not obtained from variational algorithms have been successfully used in a variety of image processing and analysis areas. Thus, both variational and non-variational PDE models are very successful in solving some important challenges which still persist in digital image processing and computer vision fields.

So, numerous PDE-based and variational techniques that solve properly various image processing and computer vision tasks, such as the image denoising and restoration, inpainting, image segmentation, image registration and the video motion analysis by using optical flow estimation, have been proposed in the last decades. The first two of the above-mentioned PDE-based image processing and analysis subdomains are approached in this work.

© Springer International Publishing AG, part of Springer Nature 2019
T. Barbu, *Novel Diffusion-Based Models for Image Restoration and Interpolation*,
Signals and Communication Technology,
https://doi.org/10.1007/978-3-319-93006-0_1

The aim of this book is to describe the image restoration and inpanting techniques using partial differential equations, but focusing mainly on the novel and recent contributions of the author in these strongly related domains. Thus, some of the most important theoretical and practical results of our research in these areas are described here.

The book is addressed to students and professionals working in area of mathematics, computer science and electrical engineering. The required mathematical background is knowledge of the partial differential equations theory and numerical analysis at an elementary level. Also, the readers of this book need a basic knowledge of the digital image processing domain.

The content of this book is composed of five chapters, this introduction representing the first one, and a section of references. The PDE-based image denoising and restoration domain is approached in the following two chapters, while the fourth chapter is devoted to the PDE-based image inpainting field. The conclusions of this work are drawn in the last chapter.

Digital image denoising represents an important image processing task, consisting of removing the electronic noise, which is a random variation of brightness or color information, from the image signal. There are various types of image noise, but the research described here focused mainly on filtering the additive 2D noise. The multiplicative noise, which is also discussed in this book, can be reduced to additive noise by applying a logarithmic operation to the observed image in the formation model.

The most common additive noise is the Gaussian noise that represents a statistical noise characterized by a probability density function (PDF) equal to that of the normal distribution and its principal source arise during image acquisition and transmission processes. Thus, the 2D Gaussian noise model has the following form:

$$p(z) = \frac{1}{\sqrt{2\pi}\sigma}e^{-(z-\mu)^2/2\sigma^2} \tag{1.1}$$

where z is the gray level, μ returns the mean of random variable z and σ is its standard deviation, while the formation model of this type of image noise is expressed as:

$$u^* = Hu + N(0, \sigma^2) \tag{1.2}$$

where the grayscale image is given as a function $u: \Omega \subseteq R^2 \to R$, u^* is the observed image, $H : \mathcal{H} \to \mathrm{K}$, with K a real Hilbert space, is a bounded linear operator (for example, a convolution) and the term $N(0, \sigma^2)$ represents the additive noise.

A denoising process has to recover the image u from the observed image u^* provided by (1.2). The search for efficient image denoising techniques still represents a valid challenge in the image processing domain. Such a restoration approach is intended to optimize the trade-off between image smoothing, detail preservation and undesired effect removal. Unfortunately, the conventional restoration methods, such as the 2D Gaussian and Average filters, generate the unintended blurring effect and cannot preserve well the edges and other image features. The PDE-based filtering models that are described in this work represent the best solution to this problem, providing a detail-preserving image restoration.

These PDE denoising models are based on linear and nonlinear diffusion processes. The linear diffusion-based image restoration domain is presented in Chap. 2. The theory of the diffusion processes and the existing restoration approaches based on linear diffusion are described in its first section.

Our main contributions in this image denoising field are presented next. Thus, two effective hyperbolic linear PDE-based filtering models developed by us are discussed in the following sections of the chapter. They have the localization property and a fast converging character. A stochastic differential equation-based restoration framework proposed by us has been also included in this second chapter, since some linear diffusion-based approaches can be derived from it. The described linear PDE denoising models outperform the classic image filters, but have their own drawbacks.

The nonlinear PDE-based image denoising and restoration models that overcome the disadvantages of those linear diffusion schemes are presented in Chap. 3. The state of the art nonlinear diffusion-based restoration techniques are surveyed in the first section. They include second-order PDE denoising models, such as the anisotropic diffusion models inspired by Perona-Malik scheme and the total variation (TV) based algorithms, and also fourth-order PDE-based models, like the isotropic diffusion-based You-Kaveh denoising scheme. Hybrid restoration methods using nonlinear diffusions are also discussed.

Our research achievements in the nonlinear diffusion-based restoration domain are detailed in the next sections of the chapter. Thus, its second section presents the most important nonlinear second-order PDE-based filtering techniques constructed by us. They include some parabolic anisotropic diffusion models for image denoising, PDE variational restoration schemes and hyperbolic PDE-based smoothing algorithms. These described approaches provide an effective additive noise removal, while preserving essential features like edges and corners, and overcoming the blurring effect.

The nonlinear fourth-order diffusion-based noise reduction approaches developed by us are described in the third section of the chapter. Several fourth-order parabolic diffusion-based schemes, modeled in PDE and variational forms, and hyperbolic PDE-based restoration models are presented in the first two subsections. These techniques represent successful denoising solutions and also solve the shortcomings of the second-order diffusion-based methods, by overcoming the undesired staircase effect.

Since the fourth-order PDE restoration models have their own shortcomings, such as generating multiplicative noise and blurring some details because of over-filtering, we have developed effective hybrid restoration solutions involving fourth-order diffusions and overcoming these drawbacks, which are discussed in the last subsection. One of them combines nonlinear second and fourth order PDEs, thus achieving a better smoothing and avoiding both blurring and staircasing. Another compound scheme combining a fourth-order diffusion model to a 2D Gaussian filter kernel and a despeckling algorithm removes successfully both white additive Gaussian noise and the speckle multiplicative noise, while overcoming the blocky effect.

We have also developed a class of variational denoising frameworks based on nonlinear control problems, which is described in the last section of the third chapter. Thus, some effective image restoration techniques are obtained by solving nonconvex optimal control problems with the state and controller connected on a manifold that is described by a nonlinear elliptic PDE.

The second image processing field approached in this book, digital image inpainting (interpolation, reconstruction), is closely related to image restoration, since an interpolation procedure reconstructs the missing part of the image by directing the denoising process to that inpainting domain. Therefore, the PDE-based filtering models can be easily adapted for the inpainting task.

As mentioned before, the nonlinear diffusion-based image reconstruction domain is addressed in Chap. 4, which is composed of three sections. The state of the art PDE-based inpainting techniques are presented in the first section. The variational image interpolation field, including influential energy-based inpainting models such as Harmonic Inpainting, Mumford-Shah based Inpainting, total variation-based schemes (TV and TV^2 Inpainting) and Euler's Elastica Inpainting, is described first. Then, the second-order and higher-order PDE inpainting techniques are presented. While the second-order diffusion-based reconstruction schemes are easily derived from variational models, the high-order PDE-based inpainting algorithms do not follow variational principles. They include third-order PDE schemes, such as the inpainting approach introduced by Bertalmio et al. and the Curvature-driven Diffusion (CDD) Inpainting, and fourth-order PDE models, such as Cahn-Hilliard Inpainting, $TV-H^{-1}$ Inpainting and LCIS Inpainting.

We have conducted a lot of research in the PDE-based structural inpainting area in the last decade. Our most important research results accomplished in this field are described in the next two sections of the chapter. The second section contains the variational structure-based image reconstruction algorithms proposed by us. A class of effective variational interpolation solutions that reconstruct properly the inpainting domain, overcome the undesirable effects and work successfully in both normal and noisy conditions is described first. Then, we provide a hybrid variational inpainting framework that combines second and fourth order nonlinear diffusion components, so that to achieve an improved inpainting and reduce the additive noise and the undesirable effects.

The nonlinear PDE inpainting models elaborated by us are detailed in the third section. Some effective parabolic second-order anisotropic diffusion-based interpolation techniques are discussed in the first subsection. Then, a class of nonlinear hyperbolic PDE-based interpolation techniques is described in the second subsection. This reconstruction solution fills in successfully the missing part of the image, performing properly in noisy conditions also. Given its hyperbolic character, it reduces the diffusion effect near boundaries, preserving them very well.

Next, an improved nonlinear anisotropic diffusion-based inpainting framework is introduced in the third subsection. It does not follow a variational principle and has a more complex character than other nonlinear diffusion approaches. Its PDE model is based on two diffusivity functions and an additional component that controls the speed of the diffusion process. This anisotropic diffusion technique inpaints successfully the deteriorated images, while reducing the white Gaussian noise and the unintended effects. Thus, it performs very well in noisy conditions and preserves the image details.

Rigorous mathematical treatments are also performed on the PDE-based image restoration and interpolation models addressed in this book. So, the edge-stopping functions of the proposed diffusion-based schemes are investigated, since they have to satisfy the requirements of a proper image diffusion. Other mathematical investigations are performed on the localization property of some PDE models, but the most important mathematical treatments presented in this book are performed on the validity of the proposed models.

So, the well-posedness of the considered PDE denoising and inpainting models is seriously treated. The most of the differential models described here are well-posed, the existence of some unique weak solutions for them being demonstrated, eventually under some certain assumptions. These solutions are then computed using some iterative numerical algorithms.

Thus, an iterative explicit numerical approximation scheme is constructed for each PDE-based model, by applying the finite-difference method. All the discretization schemes described in this book are consistent to the corresponding PDE models and stable, so they are convergent and converge fast to the solutions of those PDEs. These finite difference-based numerical algorithms have been successfully used in our many restoration and inpainting experiments that are also discussed in the book. These successful tests prove the effectiveness of the considered PDE-based denoising and interpolation methods. Their performance is assessed by using various image quality measures, the most commonly used being Peak Signal to Noise Ratio (PSNR) that has the form:

$$PSNR = 10 \log_{10} \left(\frac{\max(u)^2}{MSE} \right) \tag{1.3}$$

where $\max(u)$ returns the maximum pixel value of the original image u and

$$MSE = 1 / IJ \sum_{i=0}^{I-1} \sum_{j=0}^{J-1} [u(i, j) - u_0(i, j)]^2 \tag{1.4}$$

where u_0 is the observed $[I \times J]$ image and the Mean Squared Error (MSE) represents another image quality metric used in the experiments. Other performance estimators used here are Structural Similarity Index (SSIM) and Norm of the Error Image.

Our diffusion-based restoration and inpainting techniques achieve good values for these performance measures. Method comparison performed by us are also described in the three chapters. Our approaches are compared to many state of the art algorithms from their category, and outperform them or perform comparable well to them, according to the method comparison results.

As already mentioned, the fifth chapter is devoted to the conclusions of this book. It also discusses the future research plans and some possible application areas of the described techniques. The manuscript ends with a section of bibliographical references that lists all the works cited, many of them representing published papers authored by us.

Chapter 2
Linear PDE-Based Image Denoising Schemes

The aim of this chapter is to describe the linear diffusion-based image filtering domain. The diffusion process theory and the existing linear partial differential equation (PDE)—based denoising techniques are discussed in the first section. Then, our main contributions to this image processing field are described in the following sections. Thus, an effective digital image restoration approach using a hyperbolic second-order differential model, is detailed in the second section. Another linear dynamic partial differential equation-based scheme for image smoothing is described in the third section. The last section of this chapter presents a stochastic differential equation (SDE)—based restoration model leading to a linear diffusion approach.

2.1 Related Works

The *diffusion* represents the process in physics which is related to moving from an area of high concentration to an area of low concentration. The diffusion process equilibrates the concentration differences without creating or destroying the mass. It is expressed by the Fick's *first law of diffusion* [1], having the following form:

$$J = -D \cdot \nabla u \qquad (2.1)$$

where J represents the *diffusion flux*, D is the *diffusivity* (or *diffusion coefficient*) and ∇u is the *concentration gradient*. Since the mass is conserved in this diffusion process, which does only transport it without destroying, one may apply the *continuity equation* [1], a conservation law that describes the physical transport process:

$$\frac{\partial u}{\partial t} = -div\, J \qquad (2.2)$$

© Springer International Publishing AG, part of Springer Nature 2019   7
T. Barbu, *Novel Diffusion-Based Models for Image Restoration and Interpolation*,
Signals and Communication Technology,
https://doi.org/10.1007/978-3-319-93006-0_2

By replacing the flux J provided by Eq. (2.1) in the Eq. (2.2) one obtains the *diffusion equation*:

$$\frac{\partial u}{\partial t} = div(D \cdot \nabla u) \tag{2.3}$$

The diffusion equation expressed by Eq. (2.3) is used by numerous domains, such as the heat transfer, where is called *heat equation*, and image processing, where the concentration is identified to the grayscale value at a given location. The diffusion processes are very useful to image denoising, an important image processing sub-domain [2].

If one considers the grayscale image $u: \Omega \subseteq R^2 \rightarrow R$, the observed image $I \in L^1(R^2)$ is affected by the additive Gaussian noise given by Eq. (1.1) according to the formation model Eq. (1.2) that is $I = Hu + N(0, \sigma^2)$. Thus, the next diffusion model is obtained for the image denoising process:

$$\begin{cases} \frac{\partial u}{\partial t} = div(D(x, y, t) \cdot \nabla u) \\ u(0, x, y) = I(x, y) \end{cases}, (x, y) \in \Omega \subseteq R^2 \tag{2.4}$$

where $u(x, y, t)$ is the image obtained after a diffusion time t, Ω represents the image domain, and $D(x, y, t)$ represents the diffusivity function (also called *diffusion tensor*).

If this diffusivity does not depend on the evolving image u itself, then the partial differential Eq. (2.4) represents a *linear diffusion* process. Otherwise, if D is a function of u, we have a *nonlinear diffusion*. If the diffusivity is constant over the entire image domain ($D(x, y, t) = \alpha, \forall (x, y) \in \Omega$), then the PDE given by Eq. (2.4) represents a *homogeneous*, or *isotropic diffusion*. If D represents a space-dependent function, then the diffusion process is *inhomogeneous*, or *anisotropic* [2].

The linear diffusion procedures represent the simplest and most investigated PDE-based denoising techniques. Since we have

$$\frac{\partial u}{\partial t} = div(D(x, y, t) \cdot \nabla u) = D(x, y, t)\Delta u + \nabla D(x, y, t) \cdot \nabla u \tag{2.5}$$

we get the following linear isotropic diffusion model from (2.4):

$$\begin{cases} \frac{\partial u}{\partial t} = div(\alpha \nabla u) = \alpha \cdot \Delta u \\ u(0, x, y) = I(x, y) \end{cases}, (x, y) \in \Omega \tag{2.6}$$

where $\alpha > 0$.

The PDE model provided by Eq. (2.6) represents a heat equation, which is a well-known parabolic partial differential equation describing the distribution of the heat (temperature variation) [3]. Also, this second-order PDE can be discretized easily by using consistent and stable finite-difference based schemes. The numerical approximation of the model (2.6) would produce an iterative algorithm that evolves the noised image to the denoised version.

This linear diffusion based smoothing process is equivalent to the 2D Gaussian filtering [4]. Thus, the parabolic Eq. (2.6) is characterized by a unique solution that has the following form:

$$u(x, y, t) = \begin{cases} I(x, y), & t = 0 \\ (G_\sigma * I)(x, y), & t > 0 \end{cases} \tag{2.7}$$

where $\sigma = \sqrt{2t}$ and G_σ denotes the two-dimensional Gaussian filter kernel that is expressed as:

$$G_\sigma(x, y) = \frac{1}{2\pi\sigma^2} e^{-\frac{x^2+y^2}{2\sigma^2}} \tag{2.8}$$

The parameter $\sigma > 0$ determines the spatial size of the image details which are *removed* by this 2D filter: the bigger σ, the smoother the result and the lesser details are kept. The convolution operation in Eq. (2.7) is performed as following:

$$G(x, y) * I(x, y) = \int_x \int_y G(x - \xi, y - \eta) \cdot I(\xi, \eta) d\xi \, d\eta \tag{2.9}$$

where this convolution is efficiently computed by applying the Fast Fourier Transform (FFT).

This demonstrates that performing a linear diffusion for a time t with $\alpha = 1$ is exactly equivalent to performing the Gaussian restoration with $\sigma = \sqrt{2t}$. The 2D Gaussian filter provided by Eq. (2.8) constitutes a low-pass filter that attenuates high frequencies in a monotone way, as it results from the analysis of its behaviour in the frequency domain [4].

The PDE-based linear filtering approaches have some important advantages, such as the easyness of handling, the fast execution and the low processing time. Also, the PDE formulation in terms of a diffusion equation is much more natural and posesses a higher generalization potential than the convolution with 2D Gaussians.

The main disadvantage of the linear diffusion models is the undesired image blurring effect. The isotropic diffusion schemes perform a homogeneous denoising, blurring equally in an all directions. These PDE denoising techniques remove successfully the Gaussian noise, but often blur some important image details and features, such as boundaries and textures, since they generate diffusion across the edges.

A linear diffusion-based image denoising example is displayed in the next figure. The denoised form of the noisy image from Fig. 2.1a, which is displayed in Fig. 2.1b, is affected severely by the blurring effect.

Also, the linear PDE-based filtering approaches could dislocate the edges when moving from finer to coarser scales [5]. Another drawback is that linear diffusion

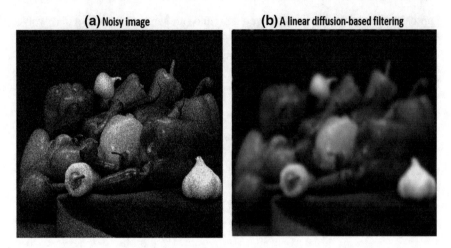

Fig. 2.1 A linear PDE-based denoising example

models expressed by Eq. (2.6) have no localization property, their solutions propagating with infinite speed.

Many existing denoising approaches try to address these disadvantages by proposing some modifications of the filter kernels or by transforming the linear PDE models into nonlinear diffusion processes. Thus, a solution to preserve the essential image features during the restoration process would be the *directed diffusion* [6].

A directed diffusion process incorporates some a priori knowledge about the image details to be preserved, into the linear partial differential equation model. Such a directed diffusion method is that proposed by R. Illner and H. Neunzert [6]. Their restoration technique provides some important background information in form of an auxiliary image, A, and is expressed by the following PDE-based model:

$$\begin{cases} \frac{\partial u}{\partial t} = A \Delta u - u \Delta A \\ u(0, x, y) = g(x, y) \end{cases}, (x, y) \in \Omega \qquad (2.10)$$

Another class of linear PDE-based denoising models is that of the *linear complex diffusion schemes*. In [7] there is proposed such a linear complex diffusion model, having the form:

$$\begin{cases} \frac{\partial u}{\partial t} = u_t = c u_{xx}, t > 0, x \in R \\ u(x; 0) = u_0 \in R, c, u \in C \end{cases} \qquad (2.11)$$

The PDE in (2.11) represents a generalization of the linear diffusion equation given by Eq. (2.6) for $c \in R$. In this case the diffusion model is well-posed for $c > 0$ [7].

We have also proposed some improved linear diffusion models that overcome the drawbacks of the existing linear PDE-based schemes. They are described in detail in the next section of this chapter.

2.2 Second-Order Hyperbolic PDE-Based Restoration Solutions

In this section and the next one we describe some of our contributions in this linear PDE-based image restoration domain. They represent some improved linear diffusion schemes that aim to solve the main drawbacks of the existing isotropic linear PDE-based denoising models.

The proposed restoration approaches are based on second-order hyperbolic equations [8–10]. Thus, we consider a class of hyperbolic diffusion-based restoration models. It is composed of a linear second-order hyperbolic partial differential equation and a set of boundary conditions [10]. So, we have:

$$
\begin{cases}
\lambda \frac{\partial^2 u}{\partial t^2} + \gamma^2 \frac{\partial u}{\partial t} - \alpha \nabla^2 u + \zeta (u - u_0) = 0 \\
u(0, x, y) = u_0(x, y) \\
\frac{\partial u}{\partial t}(0, x, y) = u_1(x, y) \\
u(t, x, y) = 0, \forall t \geq 0, (x, y) \in \partial \Omega
\end{cases}
, (x, y) \in \Omega \qquad (2.12)
$$

where $\Omega \subseteq (0, \infty) \times R^2$ represents the image domain, $\lambda, \gamma, \alpha \in (0, 3]$, $\zeta \in (0, 0.5]$ and u_0 represents the initial image, which is corrupted by Gaussian noise.

The 2nd—order dynamic PDE model expressed by Eq. (2.12) is well-posed. Assuming that $u_0 \in L^2(R^2)$, then this isotropic diffusion-based equation is having a unique weak solution that is continuous in t with values in $L^2(R^2)$. Moreover, if $u_0 \in H^k(R^2)$, then $u(t) \in H^k(R^2), \forall t \geq 0$ [9].

Also, this PDE scheme represents a non-Fourier model for the heat propagation, since its unique and weak solution is propagating with finite speed [10, 11]. Thus, one may demonstrate that if the support of u_0 is in the ball of radius r, $B_r = \{(x, y); x^2 + y^2 \leq r^2\}$, then the support of $u(t, x, y) \subseteq B_{r+\rho t}$, for some $\rho > 0$. See [12], at page 260 for more proof. That means, the proposed differential model possesses the localization property [5], unlike others linear PDEs, representing Fourier models of heat propagation and having solutions propagating with infinite speed, therefore not having this important property.

The component $\zeta (u - u_0)$ from the Eq. (2.12) is introduced to stabilize this optimal denoising solution, preventing the further degradation of the filtered image. This weak solution of the linear diffusion model is approximated numerically by using the finite-difference method.

So, a consistent explicit numerical approximation scheme is developed for the proposed PDE model [13, 14]. One considers a space grid size of h and a time step Δt. The space and time coordinates are quantized as follows:

$$x = ih, y = jh, t = n\Delta t, \forall i \in \{0, 1, \ldots, I\}, j \in \{0, 1, \ldots, J\}, n \in \{0, 1, \ldots, N\} \tag{2.13}$$

The second-order equation from (2.12) is equivalent to the second-order PDE $\lambda \frac{\partial^2 u}{\partial t^2} + \gamma^2 \frac{\partial u}{\partial t} - \alpha \left(\frac{\partial^2 u}{\partial x^2} + \frac{\partial^2 u}{\partial y^2} \right) + \zeta(u - u_0) = 0$, which can be discretized, by using the finite differences, as:

$$\lambda \frac{u^{n+\Delta t}(i,j) + u^{n-\Delta t}(i,j) - 2u^n(i,j)}{\Delta t^2} + \gamma^2 \frac{u^{n+\Delta t}(i,j) - u^{n-\Delta t}(i,j)}{2\Delta t} -$$
$$\alpha \frac{u^n(i+h,j) + u^n(i-h,j) + u^n(i,j+h) + u^n(i,j-h) - 4u^n(i,j)}{h^2} \tag{2.14}$$
$$+ \zeta \left(u^n(i,j) - u^0(i,j) \right) = 0$$

We may take $h = 1$ and $\Delta t = 1$, so (2.14) leads to the next explicit numerical approximation scheme of the differential model:

$$u^{n+1}(i,j) = \frac{2\zeta - 4\lambda}{2\lambda + \gamma^2} u^n(i,j) + \frac{\gamma^2 - 2\lambda}{2\lambda + \gamma^2} u^{n-1}(i,j) + 2\zeta u^0(i,j) +$$
$$2\alpha \left(u^n(i+1,j) + u^n(i-1,j) + u^n(i,j+1) + u^n(i,j-1) - 4u^n(i,j) \right) \tag{2.15}$$

where $u^0(i,j) = u_0(i,j)$ and $n > 0$.

The obtained iterative restoration algorithm receives a $[I \times J]$ noisy image as input and applies repeatedly the operation (2.15), for $n = 1, 2, \ldots, N$. The number of iterations of our numerical approximating scheme, N, is quite low, since this discretization procedure converges fast to the solution representing the optimal denoising.

The described linear hyperbolic diffusion technique has been successfully tested on hundreds of images corrupted by various levels of Gaussian noise. Our restoration model achieves satisfactory smoothing results while preserving the image features, such as edges and other important details, quite well, although the image blurring is not completely avoided. As one can see in the next figure, our approach overcomes other unintended effects, such as the image staircasing [15] and speckle noise [16].

We have identified the following set of PDE model's parameters that provide the optimal results:

$$\lambda = 2.4, \gamma = 1.5, \alpha = 1.8, \zeta = 0.15, \Delta t = 1, N = 12 \tag{2.16}$$

So, the optimal image smoothing is achieved after a low number of iterations, $N = 12$. That means our denoising technique runs very fast, its execution time being less than 1 s.

Table 2.1 PSNR values for several noise removal techniques

Our filter	Linear PDE-based filter	Average	Wiener	Perona-Malik	TV
27.1 (dB)	24.3 (dB)	23.2 (dB)	24.5 (dB)	25.8 (dB)	24.3 (dB)

Method comparison were also performed and we found that our restoration technique outperforms not only the 2D conventional filters, but also numerous linear and nonlinear PDE-based approaches. The smoothing performance of the proposed approach is assessed using the Peak Signal-to-Noise Ratio (PSNR) measure [17].

Our denoising scheme obtains higher PSNR values than popular 2D classic filters, such as two-dimension Gaussian, Average and Wiener [4], the linear PDE-based denoising (heat equation) and influential nonlinear diffusion-based models, like Perona-Malik scheme and TV Denoising (see more about them in the next chapter). One can see the results in Table 2.1, which registers the average PSNR values in decibels.

The denoising results produced by these techniques are displayed in Fig. 2.2, which contains: (a) the original [512 × 512] *Peppers* image; (b) the image corrupted by Gaussian noise with parameters $\mu = 0.21$ and *variance* $= 0.023$; (c) the image restored by our scheme; (d) Linear PDE-based filtering (based on heat equation); (e) and (f) the denoising results achieved by the [3 × 3] 2D filter kernels (two-dimension Average and Wiener); (g) Perona-Malik restoration; (h) TV denoising.

Therefore, our linear isotropic PDE restoration model achieves a better enhancement than conventional filters and existing linear diffusion models, removing a greater amount of Gaussian noise and reducing the blurring effect. It also provides better restoration results than some nonlinear PDE and variational models, executes much faster, by converging to the optimal solution in fewer iterations, and overcomes better other unintended effects, such as the staircasing and speckling.

While some improved second-order nonlinear PDE models achieve better filtering results (higher PSNR) than our linear hyperbolic scheme and a better deblurring, our technique possesses some advantages over them, too. It avoids totally the staircase effect and executes considerably faster than those approaches, having a lower computational cost.

Also, this proposed linear PDE-based model is important not only because it operates successfully, produces satisfactory noise reduction results and outperforms other smoothing algorithms, but also because some effective nonlinear diffusion based image enhancement schemes can be derived from it. For example, a second-order nonlinear hyperbolic PDE model that would provide much better deblurring results can be obtained by transforming the Eq. (2.12) as following:

$$\lambda \frac{\partial^2 u}{\partial t^2} + \gamma^2 \frac{\partial u}{\partial t} - \alpha(\Delta u)\Delta u + \zeta(u - u_0) = 0 \qquad (2.17)$$

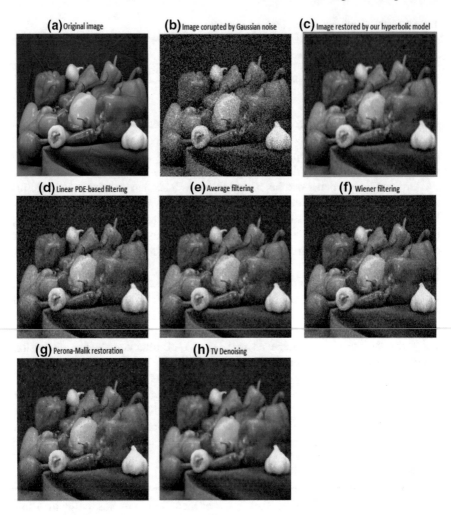

Fig. 2.2 Method comparison: image restored by various denoising techniques

where α does not represent a constant anymore, but a function of the current image's Laplacian, $\Delta u = \nabla^2 u$. Another version of (2.17) may use it as a function of the gradient, $\alpha(\nabla u)$.

We will describe both second-order and fourth-order nonlinear versions of this linear hyperbolic diffusion model in the second and third sections of the next chapter.

2.3 Linear Anisotropic Diffusion-Based Smoothing Technique

We describe here another linear dynamic PDE model for noise reduction, which is more effective than existing isotropic linear denoising schemes and overcomes the unintended effects. Unlike the PDE model described in the previous section, this diffusion-based image restoration approach has an anisotropic character [18]. So, our model is composed of a linear second-order hyperbolic PDE and a set of boundary conditions. It is expressed as follows:

$$
\begin{cases}
\alpha^2 \frac{\partial^2 u}{\partial t^2} + \beta \frac{\partial u}{\partial t} - \frac{\gamma^2}{2} \Delta u + E \cdot \nabla u = 0 \\
u(0, x, y) = u_0(x, y), (x, y) \in \Omega \\
u(t, x, y) = 0, \forall t \geq 0, (x, y) \in \partial \Omega
\end{cases}
\tag{2.18}
$$

where the domain $\Omega \subseteq (0, \infty) \times R^2$, the coefficients $\alpha, \beta, \gamma \in (0, 1]$, u_0 is the initial noisy image, while the function $E : R^2 \to R$ takes the following form:

$$
E(x, y) = \left(e^{-\eta(x^2+y^2)}, e^{-\xi(x^2+y^2)} \right)
\tag{2.19}
$$

where $\eta, \xi > 0$. Since it represents a space-dependent function, depending on the space coordinates x and y, the function E provides the anisotropic character of the model (2.18).

This second–order linear anisotropic diffusion approach represents a well-posed PDE model, since it has a unique and weak solution, u. Because this solution is propagating with finite speed [11], the proposed differential scheme constitutes a non-Fourier model for the heat propagation. That means it also has the localization property [5], unlike the most isotropic linear PDE-based denoising models.

This unique weak solution of the model (2.18) is computed by using a numerical approximation scheme. So, an explicit finite-difference based discretization scheme that is consistent to the continuous model is developed by us in [18].

As in the previous case, described in 2.2, we consider here a space grid size of h and a time step Δt, the space and time coordinate quantization being given by Eq. (2.12).

The hyperbolic diffusion-based equation provided by Eq. (2.18) leads to the next equation:

$$
\alpha^2 \frac{\partial^2 u}{\partial t^2} + \beta \frac{\partial u}{\partial t} - \frac{\gamma^2}{2} \left(\frac{\partial^2 u}{\partial x^2} + \frac{\partial^2 u}{\partial y^2} \right) + \left(e^{-\eta(x^2+y^2)}, e^{-\xi(x^2+y^2)} \right) \cdot \left(\frac{\partial u}{\partial x}, \frac{\partial u}{\partial y} \right) = 0
\tag{2.20}
$$

that is approximated by using the finite difference method [13], as:

$$\alpha^2 \frac{u^{n+\Delta t}(i,j)+u^{n-\Delta t}(i,j)-2u^n(i,j)}{\Delta t^2} + \beta \frac{u^{n+\Delta t}(i,j)-u^{n-\Delta t}(i,j)}{2\Delta t} -$$

$$\frac{\gamma^2}{2} \frac{u^n(i+h,j)+u^n(i-h,j)+u^n(i,j+h)+u^n(i,j-h)-4u^n(i,j)}{h^2} \tag{2.21}$$

$$+\left(e^{-\eta\left(\frac{i^2}{h^2}+\frac{j^2}{h^2}\right)}, e^{-\xi\left(\frac{i^2}{h^2}+\frac{j^2}{h^2}\right)}\right)\left(\frac{u^n(i+h,j)+u^n(i-h,j)}{2h}, \frac{u^n(i,j+h)+u^n(i,j-h)}{2h}\right) = 0$$

If we take $h = 1$ and $\Delta t = 1$ respectively, then (2.21) leads to the following explicit numerical approximation scheme of the PDE model:

$$u^{n+1}(i, j) = \frac{4\alpha^2}{2\alpha^2+\beta}u^n(i, j) + \frac{\beta-2\alpha^2}{2\alpha^2+\beta}u^{n-1}(i, j)+$$

$$\frac{\gamma^2}{2\alpha^2+\beta}(u^n(i+1, j) + u^n(i-1, j) + u^n(i, j+1) + u^n(i, j-1) - 4u^n(i, j))$$

$$-\frac{2}{2\alpha^2+\beta}\left(e^{-\eta(i^2+j^2)}, e^{-\xi(i^2+j^2)}\right)\left(\frac{u^n(i+1,j)+u^n(i-1,j)}{2}, \frac{u^n(i,j+1)+u^n(i,j-1)}{2}\right)$$

$$\tag{2.22}$$

This iterative process given by Eq. (2.22) starts with the initial $[I \times J]$ observed image and applies repeatedly the above operation, for each n from 1 to N. As in the previous linear PDE denoising case, the number of iterations of the scheme, N, is very low, since the discretization procedure converges fast to the solution representing the optimal restoration.

The described linear PDE-based filtering technique has been successfully tested on hundred images affected by various amounts of Gaussian noise. The proposed image restoration scheme produces satisfactory denoising results while preserving the image details, such as boundaries and corners, very well. Our approach overcomes the undesired effects, such as staircase, blurring or speckle noise.

The next set of empirically detected parameters of this diffusion-based model provides the optimal smoothing results:

$$\alpha = 0.8, \beta = 0.5, \gamma = 0.7, \eta = 2, \xi = 3, \Delta t = 1, h = 1, N = 15 \tag{2.23}$$

The optimal restoration is reached after a low number of iterations, $N = 15$, which means the proposed algorithm is running quite fast. Its execution time is around one second.

The performed method comparison proves that our linear diffusion technique outperforms the two-dimension conventional filters, the linear isotropic diffusion-based models and also some nonlinear PDE-based techniques. Its restoration performance is assessed by using the well-known Peak Signal-to-Noise Ratio (PSNR) measure.

The described PDE filtering approach produces higher average PSNR values than the 2D classic filters, like Gaussian 2D, Average, Median or Wiener [4], and some well-known nonlinear PDE and variational models, such as Perona-Malik algorithm and TV Denoising. One can see these values registered in the following table (Table 2.2).

Table 2.2 PSNR values for the noise reduction techniques

This diffusion-based model	26.81 (dB)
Gaussian 2D	22.38 (dB)
Average filter	23.17 (dB)
Median filter	23.92 (dB)
Wiener filter	24.73 (dB)
Perona-Malik scheme	25.89 (dB)
TV Denoising	25.24 (dB)

The filtering results obtained by these approaches are displayed in Fig. 2.3, where one can see: (a) the original [512×512] *Barbara* image; (b) the image degraded by a Gaussian noise of $\mu = 0.21$ and *variance* $= 0.023$; (c) the image enhanced by our PDE-based method; (d)–(g) restoration results achieved by classic [3×3] 2D filters (Gaussian, Averaging, Median and Wiener); (h) the denoising provided by Perona-Malik model; (i) TV Denoising result.

Therefore, like the previous linear PDE model proposed by us, this restoration technique provides a better noise removal than conventional filters and other diffusion-based schemes, removing a higher amount of white additive Gaussian noise and providing better image deblurring results. It also overcomes better the other undesired effects [15, 16].

Although some second-order nonlinear diffusion approaches obtain higher PSNRs than our technique and remove totally the blurring effect, sometimes our method is preferable to them, since it overcomes completely the staircasing and runs much faster (fewer iterations) than those techniques.

Also, like the previous linear model, this approach can be further modified, so that much more performant nonlinear PDE image restoration schemes be obtained. Second-order nonlinear PDE models that would provide better image denoising results could be developed by replacing the coefficient $\frac{\gamma^2}{2}$ to some functions in (2.17):

$$\alpha^2 \frac{\partial^2 u}{\partial t^2} + \beta \frac{\partial u}{\partial t} - f(\Delta u)\nabla^2 u + E \cdot \nabla u = 0 \qquad (2.24)$$

Some fourth-order nonlinear diffusion approaches can also be derived from this linear hyperbolic equation-based technique.

2.4 Stochastic Differential Models for Digital Image Filtering

Although the most mathematical models that are successfully used in image processing and analysis fields are based on partial differential equations, the stochastic differential equation (SDE)—based models are also increasingly used in these domains.

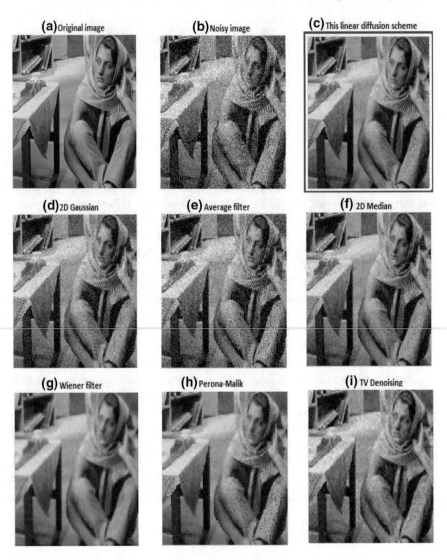

Fig. 2.3 Method comparison: image denoised by various approaches

These SDE-based schemes represent probability models that provide an effective image denoising and restoration, while preserving the image edges and other important details. The SDE-based denoising approaches are more realistic than PDE-based restoration techniques because of the presence of stochastic perturbations in images.

So, many smoothing methods based on stochastic differential equations have been elaborated in the recent years [19]. Some effective SDE image denoising techniques are based on modified diffusion [20], reflected stochastic equations [21], and stochastic relaxation and annealing [22].

Our contribution in the SDE-based restoration area is described in this section [23]. The SDE model developed by us is included in this chapter describing linear PDE models because some effective linear diffusion-based denoising schemes are derived from it.

Thus, the next stochastic differential equation-based model for additive Gaussian noise removal is introduced in [23]:

$$\begin{cases} dX(t) + F(X(t))dt = dW(t) \\ X(0, x, y) = X_0(x, y) \in R^2 \end{cases} \tag{2.25}$$

where $X(t) = \{X_1(t), X_2(t)\}$ represents the diffusion process and $W(t) = \mu\{\beta_1(t), \beta_2(t)\}$, $\mu \in (0, 1)$ is the 2D Brownian motion in the probability space $\{\Omega, F, P\}$ with the natural filtration (F_t), $t \geq 0$.

Let us assume that the function $F : R^2 \to R^2$ is Lipschitzian and set $F(X(t)) = \{F_1(X_1(t), X_2(t)), F_2(X_1(t), X_2(t))\}$. While $X(t)$ constitutes a random variable, $X_0(x, y)$ is a function on R^2, being related to the initial image.

The solution of the SDE in (2.25) is a stochastic process $X = X(t, X_0)$, adapted to the natural filtration (F_t) $t \geq 0$, on the probability space, that satisfies the next equation:

$$X(t) + \int_o^t F(X(s))ds = X_0 + W(t), t \geq 0 \tag{2.26}$$

Such a solution exists and we refer to [24, 25] for demonstrating the existence of a unique solution for the Eq. (2.26). Now, let the function $u_0 : R^2 \to R$ represent the observed image to be smoothed. Then, the restored image $u(t)$ will be determined as:

$$u(t, X_0(x, y)) = E[u_0(X(t), X_0(x, y))], t \geq 0 \tag{2.27}$$

where E returns the expectation of the argument.

We then we consider the Kolmogorov equation [26, 27] corresponding to the SDE-based model (2.25). According to [28], the recovered image represents the solution $u = u(t, \xi)$ of the linear parabolic PDE model having the following form:

$$\begin{cases} \frac{\partial u}{\partial t}(t, \xi) = \frac{\mu^2}{2}\Delta_\xi u(t, \xi) - F(\xi) \cdot \nabla_\xi u(t, \xi), t \geq 0 \\ u(0, \xi) = u_0(\xi), \xi \in R^2 \end{cases} \tag{2.28}$$

where $\xi = X_0(x, y) = \{(i, j)\}_{i=\overline{1,M}, j=\overline{1,N}} \in R^2$ and parameter $\mu \in (0, 1]$, respectively.

This parabolic equation given by Eq. (2.28) is defined on all of R^2, but an image is defined on a given domain $K \subset R^2$. So, to address this problem, one replaces (2.25) by the next reflection stochastic model [21]:

$$\begin{cases} dX(t) + F(X(t))dt + N_K(X(t))dt = dW(t) \\ X(0) = \xi \end{cases} \tag{2.29}$$

where K represents a convex subset of R^2 and $N_{;K}$ is the normal cone to the boundary ∂K.

Therefore, the function u given by Eq. (2.27) satisfies the linear partial differential Eq. (2.28) with Neumann boundary value conditions that has the following form:

$$\begin{cases} \frac{\partial u}{\partial t}(t, \xi) - \frac{\mu^2}{2} \Delta_\xi u(t, \xi) + F(\xi) \cdot \nabla_\xi u(t, \xi) = 0, \forall t \geq 0, \xi \in K \\ \frac{\partial u}{\partial v} = 0, \text{ on } \partial K \\ u(0, \xi) = u_0(\xi), \forall \xi \in K \end{cases} \tag{2.30}$$

where $u_0 : K \subseteq R^2 \to R$ represents the observed image, corrupted by Gaussian noise. Therefore, the restoration of $u = u(t, \xi)$ is provided by the solution to the linear anisotropic diffusion Eq. (2.30).

The drift term F has been properly modeled, such that to determine optimal image restoration results, when used in (2.28). Thus, we have identified the optimal form for this function, which is the following one:

$$F(X_1(t), X_2(t)) = \left(e^{-\alpha_1 \left(X_1(t)^2 + X_2(t)^2 \right)}, e^{-\alpha_2 \left(X_1(t)^2 + X_2(t)^2 \right)} \right) \tag{2.31}$$

where the coefficients $\alpha_1, \alpha_2 \geq 0$.

The well-posedness of this PDE-based scheme derived from a stochastic differential model has been rigorously investigated. We have demonstrated the existence and unicity of a weak solution for a more general form of the parabolic model given by Eq. (2.28), and under some certain assumptions in [23].

Therefore, if one assumes that F represents a Lipschitz function, then the following diffusion-based model is well-posed, admitting a unique weak solution (see [29]):

$$\begin{cases} \frac{\partial u}{\partial t} - \frac{\mu^2}{2} \Delta u + F \cdot \nabla u + g(u) = f, \text{ in } (0, T) \times R^d \\ u(0, \xi) = u_0(\xi), \forall \xi \in R^d \end{cases} \tag{2.32}$$

where $u_0 : R^d \to R$, and $f : [0, T] \times R^d \to R, d \geq 1$ and $g : R \to R$ represent two suitable chosen functions [23]. Thus, we may consider the forms $g(u) = \lambda(u - u_0)$ and $f \equiv 0$.

The solution of the PDE-based model, representing the restored image, is then approximated by using a robust numerical discretization algorithm. A consistent and fast-converging numerical approximation scheme is developed by using the finite-difference method [14].

As in the previous cases, a space grid size of h and a time step Δt and the quantization provided by Eq. (2.13) are used. The partial differential equation

$\frac{\partial u}{\partial t}(t, x, y) - \frac{\mu^2}{2}\Delta u(t, x, y) + F \cdot \nabla u(t, x, y) + \lambda(u(t, x, y) - u_0(t, x, y)) = 0$ is then discretized by using this grid. By applying the finite differences, one obtains the following discretization for that PDE:

$$\frac{u^{n+\Delta t}(i,j) - u^n(i,j)}{\Delta t} -$$

$$\frac{\mu^2}{2}\frac{u^n(i+h,j) + u^n(i-h,j) + u^n(i,j+h) + u^n(i,j-h) - 4u^n(i,j)}{h^2}$$

$$+ \left(e^{-\alpha_1\left(\frac{i^2}{h^2}+\frac{j^2}{h^2}\right)}, e^{-\alpha_2\left(\frac{i^2}{h^2}+\frac{j^2}{h^2}\right)}\right)\left(\frac{u^n(i+h,j) - u^n(i-h,j)}{2h}, \frac{u^n(i,j+h) - u^n(i,j-h)}{2h}\right) + \tag{2.33}$$

$$\lambda\left(u^n(i,j) - u^0(i,j)\right) = 0$$

If one considers the parameter values $h = \Delta t = 1$, then (2.33) leads to the next explicit numerical approximation scheme:

$$u^{n+1}(i,j) = \left(\lambda - 2\mu^2 + 1\right)u^n(i,j) +$$

$$\frac{\mu^2}{2}(u^n(i+1,j) + u^n(i-1,j) + u^n(i,j+1) + u^n(i,j-1)) \tag{2.34}$$

$$e^{-\alpha_1(i^2+j^2)}\frac{u^n(i+1,j) - u^n(i-1,j)}{2} - e^{-\alpha_2(i^2+j^2)}\frac{u^n(i,j+1) - u^n(i,j-1)}{2} - \lambda u^0(i,j)$$

The iterative numerical approximation scheme provided by Eq. (2.34) is stable and consistent to the SDE-derived differential model given by Eq. (2.28). It is also converging fast to the solution representing the denoised image, the number of iterations being quite low.

This iterative algorithm has been successfully applied on hundreds of deteriorated images, in our restoration experiments. It provides optimal restoration results when applied with the following set of parameters that are determined by the trial and error method:

$$\mu = 0.7, \alpha_1 = 2, \alpha_2 = 4, \lambda = 0.05, N = 12 \tag{2.35}$$

The performance of this SDE-based denoising technique is measured by using the Peak Signal-to-Noise Ratio (PSNR). It outperforms not only the conventional 2D filters, but also some well-known PDE-based approaches, achieving higher average PSNR values.

The proposed approach produces a better noise removal than classic filters, like Gaussian 2D, Average or 2D Wiener [4], and, unlike them, it overcomes the blurring effect and preserves much better the edges. Also, it provides a better filtering than some existing second-order PDE-based methods, such as both versions of the Perona-Malik scheme and Total Variation Denoising, removing a higher amount of white Gaussian noise, converging faster and avoiding the unintended staircase effect [15].

One can see the average PSNR values obtained by this SDE-based technique and by other PDE and non-PDE smoothing approaches, which are registered in Table 2.3.

Table 2.3 Average PSNR values achieved by some filtering methods

This model	26.94 (dB)
Gaussian	22.43 (dB)
Average	23.29 (dB)
Wiener	24.23 (dB)
Perona-Malik 1	25.69 (dB)
Perona-Malik 2	25.83 (dB)
TV Denoising	24.96 (dB)

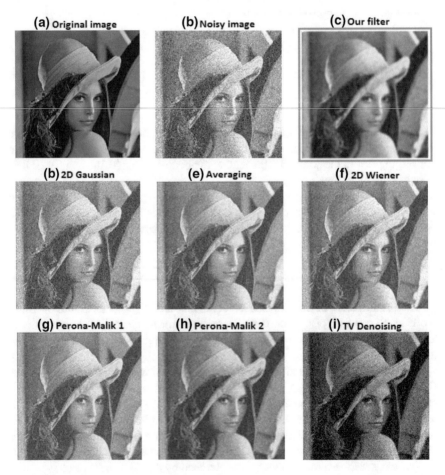

Fig. 2.4 *Lenna* image restored by various techniques

The restoration results produced by our denoising algorithm and the other filtering schemes on the *Lenna* image that is corrupted with additive 2D Gaussian noise characterized by the parameters $\mu = 0.21$ and *variance* $= 0.02$, are displayed in Fig. 2.4. These results also illustrate the smoothing performance of the proposed SDE-based method.

References

1. H. Mehrer, *Diffusion in Solids—Fundamentals, Methods, Materials, Diffusion controlled Processes* (Springer, 2007)
2. J. Weickert, *Anisotropic Diffusion in Image Processing, European Consortium for Mathematics in Industry* (B. G. Teubner, Stuttgart, Germany, 1998)
3. M.N. Ozisik, *Heat Conduction*, 2nd edn. (Wiley, New York, 1993)
4. R. Gonzalez, R. Woods, *Digital Image Processing*, 2nd edn. (Prentice Hall, 2001)
5. A.P. Witkin, Scale-space filtering, in *Proceeding of the Eighth International Joint Conference on Artificial Intelligence*, vol 2 (IJCAI '83, Karlsruhe, 8–12 August, 1983), pp. 1019–1022
6. R. Illner, H. Neunzert, Relative entropy maximization and directed diffusion equations. Math. Meth. Appl. Sci. **16**, 545–554 (1993)
7. G. Gilboa, Y.Y. Zeevi, N.A. Sochen, Complex diffusion processes for image filtering. Lect. Notes Comput. Sci. **2106**, 299–307 (2001)
8. M. Hazewinkel, *Hyperbolic Partial Differential Equation, Numerical Methods, Encyclopedia of Mathematics* (Springer Science+Business Media B.V./Kluwer Academic Publishers, 1994).
9. T. Barbu, Novel linear image denoising approach based on a modified Gaussian filter kernel, *Numer. Funct. Anal. Optim.* **33**(11), 1269–1279 (2012) (Taylor & Francis Group, LLC)
10. T. Barbu, Linear Hyperbolic Diffusion-Based Image Denoising Technique, in *Proceeding of the 22nd International Conference on Neural Information Processing, ICONIP 2015*, Part III, Istanbul, Turkey, November 9–12, vol 9491, ed. by S. Arik et al. Lecture Notes in Computer Science (Springer, 2015), pp. 471–478
11. V. Barbu, *Nonlinear Semigroups and Differential Equations in Banach Spaces* (Noordhoff International Publishing, 1976)
12. V. Barbu, *Partial Diferential Equations and Boundary Value Problems* (Kluwer Academic, Dordrecht, 1998)
13. D. Gleich, *Finite Calculus: A Tutorial for Solving Nasty Sums* (Stanford University, 2005)
14. P. Johnson, *Finite Difference for PDEs* (School of Mathematics, University of Manchester, Semester I, 2008)
15. A. Buades, B. Coll, J.M. Morel, The staircasing effect in neighborhood filters and its solution. IEEE Trans. Image Process. **15**(6), 1499–1505 (2006)
16. M. Forouzanfar, H. Abrishami-Moghaddam, Ultrasound speckle reduction in the complex wavelet domain, in *Principles of Waveform Diversity and Design*, ed. by M. Wicks, E. Mokole, S. Blunt, R. Schneible, V. Amuso (SciTech Publishing, Section B—Part V: Remote Sensing, 2010), pp. 558–77
17. E. Silva, K.A. Panetta, S.S. Agaian, Quantify similarity with measurement of enhancement by entropy, in *Proceedings: Mobile Multimedia/Image Processing for Security Applications, SPIE Security Symposium 2007*, vol 6579 (April 2007), pp. 3–14
18. T. Barbu, A linear diffusion-based image restoration approach. ROMAI Journal, ROMAI Society **2**, 133–139 (2015)
19. X. Descombes, E. Zhizhina, *Image Denoising using Stochastic Differential Equations*, RR-4814, 2003, HAL Id: inria-00071772
20. D. Borkowski, Modified diffusion to image denoising, in *Computer Recognition Systems 2, Advances in Soft Computing*, vol 45 (2007), pp 92–99

21. C. Constantini, The Skorohod oblique reflection principle in domains with corners and applications to stochastic differential equations. Probab. Theory Related Fields **91**, 43–70 (1992)
22. D. Geman, S. Geman, Stochastic relaxation. Gibbs distributions, and the Bayesian restoration of images, in *IEEE Transactions on Pattern Analysis and Machine Intelligence*, vol PAMl-6 (November 1984), pp. 721–741
23. T. Barbu, A. Favini, Novel stochastic differential model for image restoration, *Proceedings of the Romanian Academy, Series A: Mathematics, Physics, Technical Sciences, Information Science*, vol 17, num 2 (April–June 2016), pp. 109–116
24. L.C. Evans, *Introduction to stochastic differential equations* (University of California, Berkeley, 2001)
25. B. Oksendal, *Stochastic Differential Equations: An Introduction with Applications*, 3rd edn. (Springer-Verlag, New York, 1992)
26. G. Ludvigsson, *Kolmogorov Equations*, U.U.D.M. Project Report 2013:21 (Department of Mathematics, Uppsala University, 2013)
27. V. Barbu, G. Da Prato, The Kolmogorov equation for a 2D-Navier-Stokes stochastic flow in a channel. Nonlinear Anal. **69**(3), 940–949 (2008)
28. G. Da Prato, *An introduction to infinite dimensional analysis* (Springer Verlag, Berlin, 2006)
29. V. Barbu, *Nonlinear Differential Equations of Monotone Type in Banach Spaces* (Springer, 2010)

Chapter 3
Nonlinear Diffusion-Based Image Restoration Models

The nonlinear diffusion-based image denoising and restoration field is addressed in this chapter. The state of the art of this image processing domain is described in the first section. Then, our major contributions in this area are detailed in the following sections. So, the second section describes the anisotropic diffusion models for image restoration based on nonlinear second-order parabolic and hyperbolic partial differential equations, proposed by us. Nonlinear fourth-order PDE-based image noise removal techniques are discussed in the third section of this chapter. The advantages and disadvantages of the second and fourth order PDE denoising model are explained in each section. The last section is devoted to the variational image filtering approaches based on nonlinear control schemes.

3.1 State-of-the-Art Nonlinear PDE-Based Filtering Approaches

The nonlinear PDE-based techniques have been increasingly used in image restoration domain, in the last three decades, since they represent much more effective image denoising solutions than conventional filters and the filtering approaches based on linear PDE models. The state of the art nonlinear diffusion schemes for image restoration are discussed in this section.

A nonlinear diffusion model for image denoising is expressed by the relation (2.4) in Sect. 2.1, on condition that diffusion tensor D represents a function of the evolving image, u. Also, the diffusion process could be either isotropic or anisotropic.

Nonlinear PDE models overcome the main drawbacks of the linear models, such as the blurring effect and the absence of the localization property. While the linear differential schemes perform a homogeneous diffusion that could affect seriously the image details, the nonlinear PDE algorithms provide a *directional* diffusion that is degenerate along the gradient direction, having the effect of filtering the image along

© Springer International Publishing AG, part of Springer Nature 2019

T. Barbu, *Novel Diffusion-Based Models for Image Restoration and Interpolation*,
Signals and Communication Technology,
https://doi.org/10.1007/978-3-319-93006-0_3

but not across the boundaries. Thus, the edges and other important image features are preserved quite well during the nonlinear diffusion-based smoothing process.

Depending on the order of the partial differential equation, the nonlinear PDE-based schemes used for noise reduction can be divided into two main categories: second-order and fourth-order PDE models. We will describe each of them in the next subsections.

3.1.1 Nonlinear Second-Order Diffusion-Based Restoration Models

The most popular second-order nonlinear anisotropic diffusion technique is the influential restoration scheme developed by P. Perona and J. Malik in 1987 [1]. It reduces the diffusivity at those locations having a larger likelihood to represent edges, being expressed by the following parabolic PDE:

$$u_t = \frac{\partial u}{\partial t} = div\big(g\big(\|\nabla u\|^2\big) \cdot \nabla u\big) \tag{3.1}$$

Obviously, it is obtained from (2.4) by considering $D(x, y, t) = g\big(\|\nabla u(t, x, y)\|^2\big)$, where the *diffusivity*, or *edge-stopping*, function $g : [0, \infty] \to [0, \infty]$ should be monotonous decreasing and convergent to 0, for a proper restoration. Perona and Malik considered two such functions for their model, namely:

$$g(s^2) = e^{-\frac{s^2}{k^2}}; g(s^2) = \frac{1}{1 + \left(\frac{s}{k}\right)^2} \tag{3.2}$$

where $k > 0$ represents the diffusivity conductance parameter [1]. They also provided a robust finite difference-based numerical discretization algorithm for this nonlinear PDE-based model [1].

The model provides effective smoothing results while preserving the image details quite well. Some denoising results achieved by using the P-M scheme are described in the tables and figures of the previous chapter. The average PSNR values obtained by each of the two Perona-Malik versions [given by the edge-stopping functions given by (3.2)] for an image dataset are displayed in Table 2.3. The *Lenna* image restoration output provided by P-M 1 and P-M 2 are displayed in Fig. 2.4g, h. Because of its strong edge-preserving character, the Perona-Malik diffusion-based framework can be also used successfully for edge detection, outperforming the conventional edge detectors [2].

In the last 30 years there have been elaborated numerous mathematical investigations of this model and many anisotropic diffusion-based restoration techniques have been derived from the Perona-Malik scheme [3, 4]. Many of these derived models provide better filtering and edge preservation results than the original diffusion-based algorithm.

Some of these nonlinear PDE denoising schemes differ from each other by the form of the edge-stopping function. This diffusivity function g has to satisfy several conditions, such as positivity, decreasing monotony, $g(0) = 1$ and convergence to zero.

For example, the total variation diffusion model has the edge-stopping function $g(s^2) = \frac{1}{|s|}$, while its regularized form is $g(s^2) = \frac{1}{\sqrt{s^2 + \varepsilon^2}}$ [4, 5]. The Charbonnier diffusion model [6] is characterized by the following edge-stopping function:

$$g(s^2) = \left(1 + \frac{s^2}{k^2}\right)^{-1/2} \tag{3.3}$$

Weickert diffusion scheme, introduced in 1998 [4, 7], uses the next diffusivity function for denoising:

$$g(s^2) = \begin{cases} 1 - e^{-\frac{C_m}{\left(\frac{s^2}{k^2}\right)^m}}, & \text{if } |s| > 0 \\ 1, & \text{if } s = 0 \end{cases} \tag{3.4}$$

where $1 = e^{-C_m}(1 + 2C_m m)$, $m \in \{2, 3, 4\}$, $C_2 = 2.3366$, $C_3 = 2.9183$ and $C_4 = 3.3148$.

The *robust anisotropic diffusion* (RAD) scheme provided by Black et al. [8] uses the robust estimation theory to model the following edge-stopping function called *Tukey's biweight*:

$$g(s^2) = \begin{cases} \left(1 - \frac{s^2}{5k^2}\right)^2, & \text{if } \frac{s^2}{5} \leq k^2 \\ 0, & \text{if } \frac{s^2}{5} > k^2 \end{cases} \tag{3.5}$$

The anisotropic diffusion-based denoising methods may also differ in the way they choose the conductance parameter. The parameter k has to be properly selected for an effective noise removal. So, when the gradient magnitude exceeds its value, the respective edge is enhanced.

While many approaches, including the Perona-Malik scheme, use a fixed k value that can be identified empirically, some authors consider that using a single conductance value for the entire restoration process is not appropriate [9]. So, a solution is to make this parameter a function of time, $k(t)$. A high $k(0)$ value can be used at the beginning of the process, then $k(t)$ is reduced gradually, as the evolving image is filtered.

Other techniques detect automatically the conductance parameter as a function of the current state of the evolving image, u. Various noise estimation approaches can be applied for its detection [10].

So, the noise can be estimated at each iteration as the difference between the average intensities of the images processed by the morphological operations of image opening and closing [10]. In this case the diffusivity conductance is determined as:

$$k = avg(u \circ S) - avg(u \bullet S) \qquad (3.6)$$

where S is the structuring element of those mathematical morphological operations.

Another solution for the k parameter is achieved by estimating the image noise using the p-norm of that image:

$$k = \frac{\sigma \|u\|_p}{m}, \qquad (3.7)$$

where m represents the number of pixels and σ is proportional to the image average intensity [10].

Other diffusion-based techniques compute the conductance diffusivity as the *robust scale* of the analyzed image, by using some statistical measures, such as the median [8]:

$$k = \sigma_e = 1.4826 median(u)\|\nabla u\| - median(u)\|\nabla u\| \qquad (3.8)$$

Other anisotropic diffusion-based restoration approaches derived from the Perona-Malik model represent regularization attempts of that influential denoising scheme. Since Perona-Malik scheme does not admit weak solutions that are unique and stable, it has an ill-posed character. So, many solutions aiming to stabilize it have been proposed.

The implicit regularization of the Perona-Malik model can be achieved by using some properly constructed finite-difference based numerical approximation schemes [11, 12]. Other regularization solutions are independent of the numerical implementation. They introduce the regularization directly into the differential equation.

Such a regularization technique, introduced by Catte et al. in 1992 [13], uses a Gaussian-smoothed version of the evolving image. Their anisotropic diffusion model is based on the following equation:

$$u_t = div(g(\|\nabla u_\sigma\| \cdot \nabla u) \qquad (3.9)$$

where

$$u_\sigma = K_\sigma * u \qquad (3.10)$$

where $\sigma > 0$ and K_σ represents a Gaussian filter kernel. This nonlinear PDE-based model is well-posed, the existence, uniqueness and regularity of its solution being demonstrated [13].

A nonlinear diffusion model for image denoising that represents a generalization of this regularization technique was proposed by J. Kacur and K. Mikula in 1995 [14]. Another modified version of this scheme was introduced by Torkamani–Azar and Tait [15]. Their anisotropic diffusion-based approach replaces the Gaussian convolution given by (3.10) to a convolution with a symmetric exponential filter.

Another class of nonlinear diffusion-based restoration techniques is that of the PDE models based on the *mean curvature motion* (MCM). An important MCM-based restoration approach is the anisotropic diffusion scheme developed by Alvarez, Lions and Morel in 1992 [16]. They proposed a class of stable nonlinear parabolic PDE-based models of the form:

$$\begin{cases} \frac{\partial u}{\partial t} - g(|G_\sigma * \nabla u|) \|\nabla u\| div\left(\frac{\nabla u}{|\nabla u|}\right) = 0 \\ u(0, x, y) = u_0(x, y) \end{cases} \tag{3.11}$$

where G_σ is the 2D Gaussian filter kernel given by (2.8) and $g : [0, \infty] \to [0, \infty]$ is a nonincreasing real function that satisfies $\lim_{s \to \infty} g(s) = 0$.

The role of the degenerated diffusion term $\|\nabla u\| div\left(\frac{\nabla u}{|\nabla u|}\right)$ is to smooth the image u on both sides of a boundary with a minimal filtering of the edge itself, while the component $g(|G_\sigma * \nabla u|)$ is used for the enhancement of the image edges. The nonlinear diffusion-based models given by (3.11) provide an effective selective smoothing and edge detection.

Morphological anisotropic diffusion algorithms, like the one introduced by C. A. Segall and S. T. Acton in 1997 [17] are also useful in edge detection tasks. Their PDE model introduces a morphological diffusion coefficient capable of filtering the small objects while conserving the edges.

Another category of nonlinear PDE-based denoising models is that based on *complex diffusion* processes. An effective nonlinear complex anisotropic diffusion-based model proposed by Gilboa et al. in 2001 is expressed in the following form [18]:

$$u_t = \nabla \cdot (c(\text{Im}(u)) \nabla u) \tag{3.12}$$

where

$$c(\text{Im}(u)) = \frac{e^{i\theta}}{1 + \left(\left|\frac{\text{Im}(u)}{k\theta}\right|\right)^2} \tag{3.13}$$

where *Im* returns the imaginary part of the image received as argument, k is a threshold parameter and the phase angle $\theta \ll 1$.

Other second-order PDE-based denoising solutions improve the Perona-Malik model by introducing various fields in the scheme. Improved anisotropic diffusion approaches like GVF (Gradient Vector Flow)-based P-M [19], INGVF-based P-M [20], GVC (Gradient Vector Convolution)-based P–M [21] and CONVEF (Convolutional Virtual Electric Field)-based P–M models [22] have been modeled in this way.

The most nonlinear second-order PDE models follow variational principles, therefore it is very common to obtain the nonlinear diffusion-based image denoising approaches from some variational problems [4, 23]. These problems represent minimizations of some properly constructed energy functionals. Such an energy cost functional is modeled as the sum of a *regularization* component and a *fidelity* term. Thus, a generic variational framework for image restoration is based on the following minimization:

$$\min_u \{E(u) = R(u) + F(u)\} \tag{3.14}$$

where $E(u)$ represents the energy functional, the regularization term is usually given in the next form

$$R(u) = \alpha \int_\Omega \psi(\|\nabla u\|)d\Omega \tag{3.15}$$

where ψ represents the regularizer function and the fidelity term usually takes the following form:

$$F(u) = \frac{\beta}{2} \int_\Omega (u - u_0)^2 d\Omega \tag{3.16}$$

where u_0 represents the initial noisy image and $\alpha, \beta > 0$.

Numerous variational PDE models for image restoration have been developed in the last 25 years. The variational algorithms have important advantages in both theory and computation, compared to other techniques. These variational models lead to the corresponding nonlinear PDE-based schemes by determining the associated Euler-Lagrange equations [4, 23], and then applying the steepest descent method [24].

An influential variational restoration approach was developed by Rudin, Osher and Fetami in 1992 [5]. Their filtering technique, named Total Variation (TV) Denoising or ROF, is based on the minimization of the TV norm. TV Denoising model is remarkably effective at simultaneously preserving the image boundaries whilst smoothing away the noise in the flat regions. This variational restoration model has the form:

$$u_{rest} = \arg\min_{u \in L^2(\Omega)} \int_\Omega \left(\|\nabla u\| + \frac{1}{2\lambda}(u_0 - u)^2\right)d\Omega \tag{3.17}$$

where the coefficient $\lambda > 0$ and u_{rest} represents the restored image. The next second-order nonlinear diffusion-based model is obtained, by applying the Euler-Lagrange equation and the steepest descent method:

$$\frac{\partial u}{\partial t} = div\left(\frac{\nabla u}{|\nabla u|}\right) + \frac{1}{\lambda}(u - u_0) \tag{3.18}$$

This total variation regularization model removes successfully the additive white Gaussian noise (AWGN), but it can be extended to other image noise models, too. Thus, the TV Denoising scheme for Laplacian noise is based on the next minimization [25]

$$\min_{u \in BV(\Omega)} \int_{\Omega} (|\nabla u| + \lambda |u - u_0|) d\Omega \qquad (3.19)$$

where $BV(\Omega)$ is the space of bounded variation images, while the TV Denoising for Poisson noise has the form [26]:

$$u_{rest} = \arg \min_{u \in BV(\Omega)} \int_{\Omega} (|\nabla u| + \lambda (u - u_0 \log u)) d\Omega \qquad (3.20)$$

While these second-order nonlinear PDE models and variational schemes provide satisfactory detail-preserving restoration results, overcoming the image blurring effect, they still have some drawbacks. Their main disadvantage is the unintended *staircase*, or *blocky*, effect that is often generated by these denoising techniques and represents creating in image of flat regions separated by artifact boundaries [27].

So, improving the second-order diffusion-based schemes so as to avoid the staircasing has represented an important research goal in the last decades. Many variational denoising techniques that improve the classic total variation regularization model given by (3.17) have been developed in the last years. Some of them succeed to alleviate this undesired blocky effect.

The ROF Total Variation Denoising using Split Bregman employs an iterative Split Bregman algorithm to solve the TV minimization problem [28]. An improved version of the ROF scheme is the TV-l_1 restoration model that is contrast invariant and able to separate image features according to their scales [29].

Adaptive TV denoising scheme represents another improved variant of the ROF model [30]. Its regularization component operates like a TV norm at the object boundaries while approximating the l_2-norm in flat and ramp image regions so as to overcome the staircasing. The anisotropic Higher Degree Total Variation (HDTV) regularization model, proposed by Hu and Jacob in 2012 [31], minimizes successfully the staircase and ringing artifacts that are common to total variation models, by introducing novel isotropic and anisotropic higher degree TV image regularization penalties. The Generalized Total Variation (GTV) regularization for image restoration, representing a generalized pth power total variation, is another improved version of the ROF-TV model [32].

We have also developed many nonlinear second-order diffusion-based image restoration techniques, in both PDE and variational form, that deal successfully with the staircasing and other undesired effects. They are described in the second section of this chapter.

A better staircase removal solution than improving the second-order diffusion-based denoising models consists of using nonlinear fourth-order PDE schemes. These approaches are presented in the next subsection.

3.1.2 Nonlinear Fourth-Order PDE-Based Denoising Schemes

The nonlinear fourth-order partial differential equations represent an important image denoising and restoration tool that has been widely used in the last two decades, because of the drawbacks of second order PDE denoising schemes. While the second-order nonlinear diffusion-based models remove successfully the Gaussian noise, overcome the blurring and preserve well the edges and other image details, they use to approximate the observed images as step images that look *blocky* and usually generate staircase effects.

Unlike the second-order PDE and variational models, the nonlinear fourth-order PDE-based methods produce piecewise planar images that look more natural and also avoid the undesired image staircasing. They achieve comparable edge-preserving noise removal results and can be also derived from variational schemes.

A very influential nonlinear fourth-order partial differential equation-based restoration model is the isotropic diffusion-based scheme introduced by Y.-L. You and M. Kaveh in 2000 [33]. Their $L2$-curvature gradient flow method is obtained from a variational approach that is based on the minimization of the following cost functional, the restored image representing its minimum:

$$E(u) = \int_{\Omega} \left(|\nabla^2 u| \right) dxdy \qquad (3.21)$$

where $\Omega \subseteq R^2$ and f represents an increasing function, therefore its derivative satisfies $f'(\cdot) > 0$.

The result of this minimization is a planar piecewise image. Since $f(|\nabla^2 u|)$ is increasing, its global minimum is at $|\nabla^2 u = \Delta u| = 0$. Consequently, the global minimum of the functional $E(u)$ occurs when $|\Delta u| = 0, \forall(x, y) \in \Omega$. So, a planar image, which is characterized by a zero Laplacian, is the global minimum of $E(u)$ [33].

The following equation is obtained from Euler-Lagrange equation that is equivalent to (3.21):

$$\nabla^2 \left[f'(|\nabla^2 u|) \frac{\nabla^2 u}{|\nabla^2 u|} \right] = 0 \qquad (3.22)$$

that leads to

$$\nabla^2\big[g\big(|\nabla^2 u|\big)\nabla^2 u\big] = 0 \tag{3.23}$$

where the function

$$g(s) = \frac{f'(s)}{s} \tag{3.24}$$

The Eq. (3.23) is then solved by applying the gradient descent method [24], the following class of nonlinear fourth-order PDE-based models being obtained:

$$\begin{cases} \frac{\partial u}{\partial t} = -\nabla^2\big[g\big(|\nabla^2 u|\big)\nabla^2 u\big] \\ u(0, x, y) = u_0(x, y) \end{cases} \tag{3.25}$$

An iterative explicit finite-difference based numerical approximation scheme is then proposed in [33], to solve numerically the isotropic diffusion-based model (3.25). The diffusivity function used in their numerical experiments is $g(s) = \frac{1}{1+\left(\frac{s}{k}\right)^2}$ [33].

The You-Kaveh restoration technique provides effective Gaussian noise removal and overcomes successfully the staircase effect but, unfortunately, it also generates multiplicative (speckle) noise. An additional despeckling algorithm is proposed in [33] to deal with this problem.

Many fourth-order diffusion-based approaches derived from You-Kaveh model and aiming to improve it have been proposed since 2000. The GVC-based fourth-order anisotropic diffusion model for image restoration developed by Wang et al. [34] introduces the Gradient Vector Convolution (GVC) field into the Y-K scheme. First, one considers the next PDE:

$$\frac{\partial u}{\partial t} = -\nabla^2\big(c_1\Delta u - c_2 u_{\eta\eta}\big) \tag{3.26}$$

where η is the direction of the gradient and the diffusion coefficients take the following forms:

$$c_1(|\nabla u|) = \frac{1}{1 + \left(\frac{|\nabla u|}{k_1}\right)^2} \tag{3.27}$$

and

$$c_2(|\nabla u|) = m \cdot |\nabla u| \cdot e^{-\left(\frac{|\nabla u|}{k_2}\right)^2}, \quad m \in (0, 1] \tag{3.28}$$

Then, $u_{\eta\eta}$ is replaced to $V_{GVC} \cdot N$, where $N = \frac{\nabla u}{|\nabla u|}$, therefore (3.26) leads to:

$$\frac{\partial u}{\partial t} = -\Delta(c_1\Delta u - c_2 V_{GVC} \cdot N) \tag{3.27}$$

The obtained nonlinear PDE-based denoising model improves the edge and texture preserving, given the outstanding ability of the GVC field of detecting boundaries. This fourth-order anisotropic diffusion model provides also a much more effective noise removal, the PSNR values achieved by it being much improved [34]. The numerical stability has been also improved by this anisotropic diffusion method.

Another state of the art nonlinear fourth-order PDE-based denoising framework is the LLT model, introduced by M. Lysaker, A. Lundervold and X. C. Tai in 2003 [35]. In this model two different functions have been considered to measure the oscillations in the noisy data. Their PDE variational scheme has the following form:

$$u = \arg \min_{u} \int_{\Omega} \left(\left| D^2 u \right| + \frac{\lambda}{2} (u - u_0)^2 \right) d\Omega \tag{3.28}$$

where $\lambda \geq 0$ and

$$\left| D^2 u \right| = \begin{cases} \left| u_{xx} \right| + \left| u_{yy} \right|, \ or \\ \left(u_{xx}^2 + u_{xy}^2 + u_{yx}^2 + u_{yy}^2 \right)^{1/2} \end{cases} \tag{3.29}$$

The LLT denoising framework outperforms the You-Kaveh scheme for an appropriate choice of λ. This fourth-order diffusion model has been tested on a broad range of medical magnetic resonance images [35]. It provides an effective feature-preserving noise removal and overcomes successfully the staircasing effect.

Another improved fourth-order PDE denoising model is that introduced by Chan et al. [36]. It has been constructed by adding a nonlinear fourth-order diffusion term to the Euler-Lagrange equations of the TV Denosing model. The restoration method reduces substantially the undesirable staircase effect, while preserving the image edges.

The fourth-order variational models with *piecewise planarity conditions* (PPCs) represent another class of improved nonlinear fourth-order PDE-based schemes. In [37] S. Kim and H. Lim introduce and analyze piecewise planarity conditions with which unconstrained fourth-order variational schemes in continuum converge to a piecewise planar image. These fourth-order PDE variational approaches holding the PPCs can restore images better than other fourth-order diffusion-based models.

We have also developed numerous nonlinear fourth-order PDE-based restoration techniques that provide effective image denoising, preserve the essential features and deal successfully with the unintended effects. These models are described in the Sect. 3.3.

Although the nonlinear fourth-order diffusion schemes remove successfully the additive Gaussian noise and avoid the undesired staircase effect, providing more natural planar images than the second-order PDE denoising models, they also have their own disadvantages. Besides generating the undesirable multiplicative noise, the fourth-order PDE-based models could produce some blurring effect, by over-filtering the processed image. Since the fourth-order diffusion-based methods damp the high frequency components of the images much faster, they could over-smooth the step boundaries.

The restoration results produced by the most influential nonlinear second and fourth order PDE denoising models on a deteriorated image are displayed in Fig. 3.1. The original *Lenna* image is displayed in (a), while the image corrupted by an amount of Gaussian noise is depicted in (b). The next images, representing the restoration results produced by Perona-Malik (c), ROF-TV (d) and You-Kaveh (e) filtering models, illustrate the strengths and weaknesses of these PDE schemes. One can observe the blocky effect that is produced by second-order PDE-based models (P-M and TV), but it is avoided by the fourth-order Y-K scheme, which produces some edge over-filtering instead.

So, despites the improvements that have been proposed in the last years, both nonlinear second and fourth order diffusion-based filtering methods may still generate some unintended effects. Therefore, an obvious solution to this problem is to develop restoration techniques that enjoy the advantages of both types of PDE models. These techniques represent hybrid (compound) restoration approaches that combine properly second and fourth diffusion-based components, so as to achieve better denoising results and avoid the undesired effects.

Such a hybrid image restoration approach that combines nonlinear second- and fourth-order PDE filters is proposed by T. Liu and Z. Xiang in [38]. They consider a second-order Perona-Malik type model and a fourth-order You-Kaveh type scheme and discretize them by using finite differences, obtaining two stable iterative numerical approximation schemes. Then, a convex combination of the two numerical approximations is performed. The hybrid method removes successfully the Gaussian noise, while avoiding the staircasing, speckle noise and edge blurring effect.

Another hybrid PDE-based denoising solution combines a second-order total variation filter to the well-known fourth-order LLT model [39]. The compound restoration technique takes the advantage of both PDE models. It preserves very well the edges and other important image details while overcoming the blocky effect in smooth zones.

An effective compound restoration framework that combines a TV Minimization (TVM) to a nonlinear fourth-order PDE model is proposed in [40]. That hybrid diffusion-based technique is based on the following convex combination of two nonlinear PDEs of different orders:

$$w = \theta u + (1 - \theta)v, \quad \theta \in [0, 1] \tag{3.30}$$

where

$$\begin{cases} u_t = \nabla \cdot \left(\frac{\nabla u}{|\nabla u|}\right) - \lambda_1(u - u_0) \\ and \\ v_t = -\left(\frac{v_{xx}}{|D^2 v|}\right)_{xx} - \left(\frac{v_{xy}}{|D^2 v|}\right)_{yx} - \left(\frac{v_{yx}}{|D^2 v|}\right)_{xy} - \left(\frac{v_{yy}}{|D^2 v|}\right)_{yy} - \lambda_2(v - u_0) \end{cases} \tag{3.31}$$

Fig. 3.1 P-M, TV-ROF and Y-K filtering results on *Lenna* image

where $|D^2 v| = \left(v_{xx}^2 + v_{xy}^2 + v_{yx}^2 + v_{yy}^2\right)^{1/2}$ and the initial conditions are $u(0, x, y) = u_0(x, y)$ and $v(0, x, y) = v_0(x, y)$ [40].

Also, the parameter θ should be a weighting function that can be found adaptively. This combined denoising approach outperforms each of its two components, avoids successfully the staircase and blurring effects, and also the multiplicative noise. Other hybrid regularization techniques combining total variation models to fourth-order PDE schemes are described in [41].

Besides the hybrid denoising solutions involving second-order diffusions, some hybrid restoration schemes combining fourth-order PDE models of different types or mixing fourth-order diffusions with non-PDE filters, have been constructed. A recently developed hybrid PDE-based restoration approach that combines two fourth-order differential models is presented in [42].

The new denoising framework is based on a combination of a mean curvature motion-based model and a nonlinear fourth-order PDE-based algorithm. It restores successfully the mixed noise corrupted images, dealing with mixture of additive Gaussian and impulse noises [42]. The proposed hybrid PDE model has the form:

$$
\begin{cases}
u_t = -|\nabla u|^\alpha \Delta \cdot \left(\frac{\Delta u}{|\Delta u|^{1+\omega}}\right) - \beta(u - u_0), & (x, y) \in \Omega \\
u(x, y, 0) = u_0(x, y), & (x, y) \in \Omega \\
\frac{\partial u}{\partial n} = 0, & (x, y) \in \partial\Omega
\end{cases}
\tag{3.32}
$$

It becomes a traditional MCM model or a fourth-order diffusion scheme, depending on the selection of the parameters $\alpha, \beta, \omega \geq 0$. Besides handling effective the mixed noises, this fourth-order hybrid MCM-based restoration technique overcomes the staircase artifacts and removes successfully the speckle noise.

Another hybrid restoration algorithm improves a nonlinear fourth-order PDE-based model by combining it to a non-PDE filtering technique. In [43], a fourth-order diffusion-based model from the You-Kaveh class is discretized and its iterative explicit numerical approximation scheme is then combined to a *relaxed median filter* that represents a modified and improved version of the classical median filter [44].

The hybrid filtering approach enjoys the benefit of both denoising methods. It preserves the fine image details and sharp corners, because of the median filtering component, and do not introduce any staircase effects, given its fourth-order diffusion component.

We have also constructed some hybrid image denoising and restoration techniques based on nonlinear PDE-based models. They are detailed later in this chapter.

3.2 Second-Order Anisotropic Diffusion-Based Image Denoising

We have widely investigated the nonlinear PDE-based image denoising and restoration domain, and developed numerous novel second and fourth order diffusion-based filtering techniques in the last ten years. In this section we will describe the most important of our contributions in the second-order diffusion-based restoration field.

The several nonlinear second-order PDE-based image denoising approaches that are presented in the following subsections have been disseminated in some works published in recognized international journals or volumes of important international conferences. We describe image restoration models that are constructed in PDE or variational form.

While the most existing second-order PDE-based denoising models have a parabolic character, we have also elaborated effective restoration solutions based on hyperbolic PDE models. Robust finite-difference based numerical approximation schemes that are consistent to these parabolic and hyperbolic PDE-based models have been also constructed.

Rigorous mathematical investigations of the proposed differential models are also provided. The well-posedness of these nonlinear diffusion-based schemes is seriously treated, the existence and uniqueness of (weak) solutions being investigated. Some regularization processes are also performed for the ill-posed PDE models. The results of the restoration experiments and method comparison are also described in the next subsections. The developed denoising techniques are compared to some PDE and variational restoration approaches from the state of the art of this domain, described in the previous section.

Some of our restoration approaches based on parabolic anisotropic diffusion models derived from Perona-Malik scheme and aiming to improve it, are addressed in the first subsection. Effective variational image restoration schemes developed by us are presented in the third subsection. Then, our nonlinear hyperbolic diffusion-based image filtering techniques will be described in the third subsection.

3.2.1 Parabolic PDE-Based Image Filtering Models

We have proposed many nonlinear parabolic anisotropic diffusion models for image restoration that improve the influential Perona-Malik denoising scheme [1]. Thus, in [45] we consider an effective detail-preserving noise removal technique based on the following parabolic PDE model:

$$
\begin{cases}
\frac{\partial u}{\partial t} = div\big(\psi_{K(u)}(|\nabla u|^2) \cdot \nabla u\big) \\
u(0, x, y) = u_0 \\
\nabla u \cdot v = 0, \text{ on } (0, T) \times \partial \Omega
\end{cases}
\quad , \quad (x, y) \in \Omega \qquad (3.33)
$$

where u_0 is the observed image, the image domain is $\Omega \subset R^2$ and v is the normal to $\partial \Omega$. The diffusivity (edge-stopping) function is $\psi_{K(u)} : [0, \infty] \rightarrow [0, \infty]$, having the form:

$$
\psi_{K(u)}(s^2) = \begin{cases} \alpha\sqrt{\frac{K(u)}{\beta s^2 + \eta}}, & \text{if } s > 0 \\ 1, & \text{if } s = 0 \end{cases} \qquad (3.34)
$$

where the parameters $\alpha, \beta \in [0.5, 0.8]$ and $\mu \in [0.5, 1)$. The conductance parameter of this function is based on some statistics of the evolving image, being constructed as:

$$K(u) = \|u\|_F \frac{median(u)}{n(u)\varepsilon} \tag{3.35}$$

where $\varepsilon \in (0, 1]$, $\|u\|_F$ is the Frobenius norm of u, $median(u)$ represents its median value and $n(u)$ is the number of its pixels.

A robust numerical approximation scheme is then proposed for this anisotropic diffusion model. It is based on a 4-nearest-neighbours discretization of the image's Laplacian, having the form:

$$u^{t+1} = u^t + \lambda \sum_{q \in N(p)} \psi_{K(u)}\left(\left|\nabla u^{p,q}(t)\right|^2\right)\left|\nabla u^{p,q}(t)\right| \tag{3.36}$$

where $\lambda \in (0, 1)$, $N(p)$ is the 4-neighborhood of pixel $p = (x, y)$, and the image gradient magnitude in a particular direction at the iteration t is computed as following:

$$\nabla u^{p,q}(t) = u(q, t) - u(p, t) \tag{3.37}$$

The iterative algorithm (3.36) is applied on the evolving image for $t = 0, 1, \ldots, N$, where N corresponds to the best restoration. The explicit numerical approximation scheme is consistent to the PDE model (3.33) and converges fast, so N has a quite low value.

A rigorous mathematical treatment is performed on this anisotropic diffusion model in [45]. First, we demonstrate that $\psi_{K(u)}$ is properly constructed, satisfying the main properties of an effective edge-stopping function [1, 3].

Obviously, we have $\psi_{K(u)}(0) = 1$. This function is always positive, since $\alpha\sqrt{\frac{K(u)}{\beta s^2 + \eta}} > 0$, $\forall s \in R$. Also, it is monotonically decreasing, because $\psi_{K(u)}(s_1^2) = \alpha\sqrt{\frac{K(u)}{\beta s_1^2 + \eta}} \leq \alpha\sqrt{\frac{K(u)}{\beta s_1^2 + \eta}} = \psi_{K(u)}(s_2^2)$, $\forall s_1 \geq s_2$. Also, $\lim_{s \to \infty} \psi_{K(u)}(s^2) = 0$.

Since the flux function $\phi'(s) = s\psi_{K(u)}(s^2)$ is monotonic increasing, because $\phi'(s) = \psi_{K(u)}(s^2) + 2s^2\psi'_{K(u)}(s^2) \geq 0$, the proposed PDE denoising scheme represents a forward parabolic equation. The existence and uniqueness of the solution of our model requires a serious mathematical investigation. It should be said that, in general, the problem (3.33) is ill-posed. It does not have a classical solution but it has a solution in weak sense that is in the sense of distributions [45].

One can prove the existence and uniqueness of a weak solution under some certain assumptions. Thus, we demonstrate that our nonlinear diffusion-based model converges if $\gamma = a^2$. So, the next modification of the function $\psi_{K(u)}$ is considered:

$$\psi_{K(u)}(s^2) = \begin{cases} \alpha \sqrt{\frac{K(u)}{\beta s^2 + \eta}}, & \text{if } s \in (0, M] \\ \frac{\alpha}{\sqrt{\gamma}}, & \text{if } s = 0 \end{cases} \tag{3.38}$$

where $M > 0$ is arbitrarily large and fixed. The function K is Lipschitzian and positive: $|K(u) - K(v)| \leq l|u - v|$ and $K(u) \geq \rho, \forall u$.

By weak solution to Eq. (3.33) we mean a function $u : (0, T) \times \Omega \to R$ such that $u \in L^2(0, T; H^1(\Omega)) \cap L^2((0, T) \times \Omega)$, $\frac{\partial u}{\partial t} \in L^2(0, T; (H^1(\Omega))')$ and for $K(u) \geq \rho, \forall u, \forall \varphi \in H^1(\Omega), t \in [0, T]$, we have:

$$\begin{cases} \frac{\partial}{\partial t} u(t, x, y)\varphi(x, y)dxdy = -\int_\Omega \psi_{K(u)}(|\nabla u(t, x, y)|^2)\nabla u(t, x, y)\nabla \varphi(x, y)dxdy \\ u(0, x, y) = u_0(x, y), \quad \forall(x, y) \in \Omega \end{cases}$$

$$\tag{3.39}$$

where $L^2(\Omega)$ is the space of all Lebesgue square integrable functions on Ω and the Sobolev space $H^1(\Omega) = \left\{ u \in L^2(\Omega); \frac{\partial u}{\partial x_i} \in L^2(\Omega), i = 1, 2 \right\}$ where $\frac{\partial u}{\partial x_i}$ is taken in the sense of distributions. In [45] we demonstrate rigorously that anisotropic diffusion model provided by (3.33) and (3.38) admits a unique weak solution of the form given by (3.39).

The presented diffusion-based denoising method has been tested on hundreds images affected by various amounts of Gaussian noise, satisfactory results being achieved. The following PDE model's parameters have provided the optimal smoothing results: $\alpha = 0.7$, $\beta = 0.65$, $\eta = 0.5$, $\varepsilon = 0.3$, $\lambda = 0.33$ and $N = 15$. Since $\eta \cong \alpha^2$, this diffusion scheme has a unique solution and converges fast to it, the N value being quite low.

Method comparison have also been performed. The proposed anisotropic diffusion-based approach outperforms many other restoration techniques, achieving better noise removal results and converging considerably faster than some PDE-based algorithms, including the Perona-Malik scheme [1] and the Total Variation (TV)-based techniques [5]. It also produces a much better filtering than the non-PDE based smoothing approaches [2].

See an image restoration example in Fig. 3.2, where there are displayed: (a) the original [512 × 512] *Peppers* image; (b) the image corrupted with Gaussian noise characterized by the parameters $\mu = 0.21$ and $var = 0.02$; (c) the image processed by the proposed AD (anisotropic diffusion) technique; (d) the Perona-Malik filtering; (e) the TV regularization result; (f)–(i) the smoothing results produced by [3 × 3] 2D Gaussian, average, median and Wiener filters [2]. The performance of our restoration technique has been assessed by using the *norm of the error image* measure, computed as $\sqrt{\sum_{x=1}^{X} \sum_{y=1}^{Y} |u^N(x, y) - u_0(x, y)|^2}$. The average NE image values are registered in Table 3.1. As one can see in this table, our anisotropic diffusion model corresponds to the lowest NE value that indicates the best image restoration.

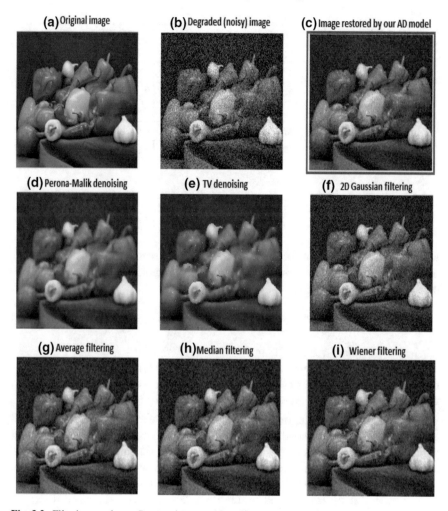

Fig. 3.2 Filtering results on *Peppers* image achieved by several models

Another parabolic second-order anisotropic diffusion-based denoising approach developed by us is described in [46]. The considered PDE model has the form:

$$\begin{cases} \frac{\partial u}{\partial t} = div(\psi_u(\|\nabla u\|)\nabla u) - v(u - u_0) \\ u(0, x, y) = u_0 \end{cases}, \quad (x, y) \in \Omega \qquad (3.40)$$

where u_0 is the initial image, its domain $\Omega \subset R^2$ and $v \in (0, 1)$. The following diffusivity (edge-stopping) function $\psi_u : [0, \infty] \rightarrow [0, \infty]$ is proposed for this model:

Table 3.1 Norm-of-the-error measure values for various denoising algorithms	Our AD model	5.15×10^3
	Perona-Malik	6.1×10^3
	TV Denoising	5.8×10^3
	2D Gaussian	7.3×10^3
	Average filter	6.4×10^3
	Median filter	6×10^3
	Wiener 2D	5.9×10^3

$$\psi_u(s) = \frac{\lambda}{\left(\frac{s}{K(u)}\right)^2 + K(u)\left|\log_{10}\left(\frac{s}{K(u)}\right)\right|} \tag{3.41}$$

where $\lambda > 1$ and the conductance parameter depends on the state of the evolving image at time t. A statistics-based automatic computation of this parameter is performed, by using the noise estimation at each time:

$$K(u(x, y, t)) = \xi\mu(\|\nabla u\|) + \alpha t, a \tag{3.42}$$

where $\xi \in (2, 3)$ and $\alpha \in (0, 1)$.

The proposed edge-stopping function ψ_u is properly selected, satisfying the conditions required by a successfully diffusion process. It is always positive and monotonically decreasing: Also, it converges to zero, since we have $\lim_{s \to \infty} \psi_u(s) = 0$.

One can prove the existence and uniqueness of a weak solution for this PDE model under some certain conditions. The proposed PDE model constitutes a forward parabolic equation that is stable and quite likely to have a solution if the flux function $s\psi_u(s)$ is monotonically increasing, which means that its derivative has to be positive. This condition, $\psi_u(s) + s\frac{\partial\psi_u(s)}{\partial s} \geq 0$, is verified for $s \leq K(u) \geq \ln(10)$.

The nonlinear diffusion-based restoration model is then discretized using the finite difference method [12]. An explicit fast-converging numerical approximation scheme that is consistent to (3.40) is developed. It is then successfully applied on many images in our restoration experiments with the next parameters that provide optimal results: $\lambda = 1.4$, $\varepsilon = 0.3$, $\xi = 2.3$, $\alpha = 0.03$, $\nu = 0.05$, $N = 14a$.

It removes the Gaussian noise and overcomes the image blurring, while preserving the image features. The performance of this restoration technique has been assessed using similarity metrics, such as PSNR (Peak Signal to Noise Ratio), and SSIM (Structural Similarity Image Metric).

See some method comparison results in Table 3.2, which registers SSIM values, and Fig. 3.3 that displays the smoothing results produced by some PDE and non-PDE filters on the *Baboon* image corrupted by additive noise. Our approach achieves higher SSIM values and better denoising results than some well-known 2D conventional filters and second-order PDE and variational models.

Table 3.2 Average SSIM values of several denoising algorithms

Method	This scheme	Average	Gaussian 2D	Perona-Malik 1	Perona-Malik 2	TV Denoising
SSIM	0.6391	0.5091	0.5412	0.6018	0.6003	0.5918

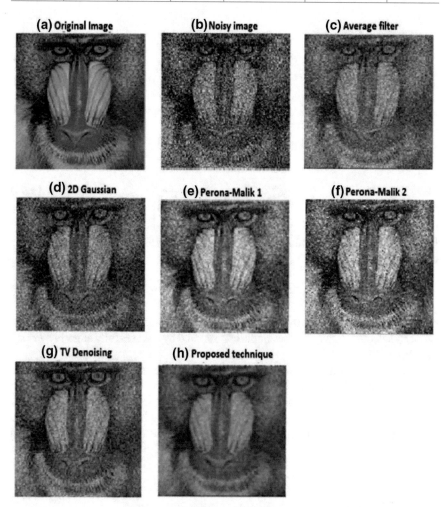

Fig. 3.3 Output of several PDE and non-PDE smoothing models

In [47] we consider another nonlinear anisotropic diffusion model that produces an effective noise removal, while preserving successfully the edges and other essential image features. It is characterized by the following PDE and boundary conditions:

Table 3.3 PSNR values obtained by some PDE and conventional filters

Model	This AD model	Average	Gaussian 2D	Perona-Malik	TV Denoising
PSNR (dB)	27.61	25.13	25.07	27.23	27.11

$$
\begin{cases}
\frac{\partial u}{\partial t} = div(\xi_u(\|\nabla u\|)\nabla u) - \lambda(u - u_0) \\
u(0, x, y) = u_0 \\
u(t, x, y) = 0, \forall t \geq 0, (x, y) \in \partial\Omega
\end{cases}
\quad , \quad (x, y) \in \Omega \quad\quad (3.43)
$$

where u_0 is the observed image, $\Omega \subset R^2$, $\lambda \in (0, 1)$ and $\partial\Omega$ is the frontier of the image domain.

The following edge-stopping function, $\xi_u : [0, \infty] \rightarrow [0, \infty]$, is proposed for this model:

$$
\xi_u(s) = \frac{\alpha}{\left(\frac{s}{\eta_u}\right)^k + \eta_u\left|\ln\left(\frac{s}{\eta_u}\right)\right|}
\quad\quad (3.44)
$$

where the parameter of conductance depends on some statistics of the evolving image u at time t, as follows:

$$
\eta_u = \beta\mu(\|\nabla u\|) + \gamma|median(u)| \quad\quad (3.45)
$$

where $\beta \in (2, 3)$ and $\gamma \in (0, 1)$. Obviously, the considered diffusivity function ξ_u is properly selected, satisfying the conditions required by a succesfully edge-preserving denoising [47].

A consistent finite difference-based discretization is then performed on this parabolic restoration model. The obtained iterative fast-converging explicit numerical approximation scheme is applied on hundreds of images corrupted by white additive Gaussian noise, satisfactory denoising results being achieved. See some method comparison results in Table 3.3, containing the average PSNR values, and in Fig. 3.4.

3.2.2 Variational Image Restoration Techniques

We have also obtained other second-order parabolic PDE-based restoration approaches from some variational models. The most important class of variational PDE image restoration schemes achieved by us will be described in this subsection.

So, in [48] we propose a class of PDE variational denoising frameworks based on the following minimization of an energy cost functional:

$$
u_{optim} = \arg \min_u E(u) \quad\quad (3.46)
$$

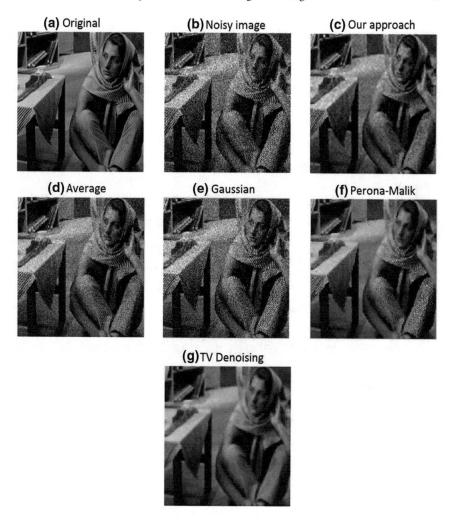

Fig. 3.4 Denoising results produced by several methods on *Barbara* image

where u_{optim} represents the restored image and

$$E(u) = \int_{\Omega} \left(\frac{\lambda}{2} \psi_u(\|\nabla u\|) + \frac{\rho}{2}(u - u_0)^2 \right) d\Omega, \qquad (3.47)$$

where the domain $\Omega \subset R^2$, the parameters $\lambda, \rho \in (0, 1)$ and u_0 is the observed image, corrupted by white Gaussian noise [48]. We propose the next regularizer function of the energy functional:

$$\psi_u(s) = \int_0^s \tau \zeta \left(\frac{\gamma(u)}{\beta \ln(s + \gamma(u))^3 + \delta} \right) d\tau, \tag{3.48}$$

whose conductance parameter is modeled as a function of some features of the evolving image, as:

$$\gamma(u) = \alpha \mu(\|\nabla u\|) + \eta t(u) \tag{3.49}$$

where $\alpha, \beta, \zeta, \eta, \delta \in (0, 3]$ and $t(u)$ returns the time of u in the evolving image sequence.

We denote $L(x, y, u, u_x, u_y) = \frac{\lambda \psi_u(\|\nabla u\|) + \rho(u - u_0)^2}{2}$, then determine the Euler-Lagrange equation corresponding to the considered variational problem [23], which is:

$$\frac{\partial L}{\partial u} - \frac{\partial}{\partial x} \frac{\partial L}{\partial u_x} - \frac{\partial}{\partial y} \frac{\partial L}{\partial u_y} = 0 \tag{3.50}$$

This leads to the following equation:

$$\rho(u - u_0) - \frac{\partial}{\partial x} \left(\frac{\lambda}{2} \psi_u'(\nabla u) \frac{2u_x}{\nabla u} \right) - \frac{\partial}{\partial y} \left(\frac{\lambda}{2} \psi_u'(\nabla u) \frac{2u_y}{\nabla u} \right) = 0 \tag{3.51}$$

that is equivalent to

$$\rho(u - u_0) - \lambda div \left(\frac{\psi_u'(\|\nabla u\|)}{\|\nabla u\|} \nabla u \right) = 0 \tag{3.52}$$

If we note $\xi_u(s) = \frac{\psi'(s)}{s}$, then (3.50) becomes:

$$\rho(u - u_0) - \lambda div(\xi_u(\|\nabla u\|)(\|\nabla u\|)\nabla u) = 0 \tag{3.53}$$

The steepest descent method is next applied on this partial differential equation. Then, by adding some boundary conditions, the following nonlinear anisotropic diffusion model is obtained:

$$\begin{cases} \frac{\partial u}{\partial t} = \lambda div(\xi_u(\|\nabla u\|)\nabla u) - \rho(u - u_0) \\ u(0, x, y) = u_0 \\ u(t, x, y) = 0, \forall t \geq 0, (x, y) \in \partial\Omega \end{cases}, \quad (x, y) \in \Omega \tag{3.54}$$

The edge-stopping function of the parabolic PDE-based model (3.54) is ξ_u : $[0, \infty] \to [0, \infty]$, constructed as following:

$$\xi_u(s) = \frac{\psi'_u(s)}{s} = \zeta \sqrt{\frac{\gamma(u)}{\beta \ln(s + \gamma(u))^3 + \delta}} \tag{3.55}$$

The function given by (3.55) is always positive, since $\xi_u(s) > 0, \forall s \geq 0$. Also, it represents a monotonically decreasing function, because $\xi_u(s_1) = \zeta \sqrt{\frac{\gamma(u)}{\beta \ln(s_1 + \gamma(u))^3 + \delta}} \geq \zeta \sqrt{\frac{\gamma(u)}{\beta \ln(s_2 + \gamma(u))^3 + \delta}} = \xi_u(s_2)$ for $\forall s_1 \geq s_2$. This diffusivity function is also convergent to zero: $\lim_{s \to \infty} \xi_u(s) = \lim_{s \to \infty} \zeta \sqrt{\frac{\gamma(u)}{\beta \ln(s + \gamma(u))^3 + \delta}} = 0$. Because it satisfies these conditions, the edge-stopping function is appropriate for an effective diffusion-based restoration.

The nonlinear anisotropic diffusion model (3.54) has also a well-posed character. A rigorous mathematical investigation of the validity of this parabolic PDE-based model is provided in [48]. The existence, uniqueness and regularity of a weak solution has been clearly demonstrated for it [48].

The PDE model is then numerically solved by constructing a robust iterative discretization algorithm that approximates that solution. The same space grid of size h and the time step Δt and the quantization (2.13) are used in this case, too. The finite difference method is then applied on the Eq. (3.54). So, since we have:

$$div(\xi_u(\|\nabla u\|)\nabla u) = \xi_u(\|\nabla u\|)\Delta u + \nabla(\xi_u(\|\nabla u\|)) \cdot \nabla u \tag{3.56}$$

a finite difference-based discretization is performed for each component of this sum [23, 48].

The first one is approximated by using the discrete Laplacian operator. Therefore, one computes $\xi_{i,j}^n = \xi_u(\|\nabla u\|_{i,j}^n)\Delta u_{i,j}^n$ for $n \in \{0, \ldots, N\}$, where

$$\Delta u_{i,j}^n = \frac{u_{i+h,j}^n + u_{i-h,j}^n + u_{i,j+h}^n + u_{i,j-h}^n - 4u_{i,j}^n}{h^2} \tag{3.57}$$

and

$$\xi_u(\|\nabla u_{i,j}^n\|) = \xi_u\left(\sqrt{\frac{\left(u_{i+h,j}^n - u_{i-h,j}^n\right)^2}{4h^2} + \frac{\left(u_{i,j+h}^n - u_{i,j-h}^n\right)^2}{4h^2}}\right) \tag{3.58}$$

The second component of the sum is computed as follows:

$$\nabla(\xi_u(\|\nabla u\|)) \cdot \nabla u$$

$$= \left(\frac{\partial}{\partial x}\xi_u\left(\sqrt{\left(\frac{\partial u}{\partial x}\right)^2 + \left(\frac{\partial u}{\partial y}\right)^2}\right), \frac{\partial}{\partial y}\xi_u\left(\sqrt{\left(\frac{\partial u}{\partial x}\right)^2 + \left(\frac{\partial u}{\partial y}\right)^2}\right)\right)$$

$$\cdot \left(\frac{\partial u}{\partial x}, \frac{\partial u}{\partial y}\right) \tag{3.59}$$

that leads to

$$\nabla(\xi_u(\|\nabla u\|)) \cdot \nabla u = \xi_u'(\|\nabla u\|) \frac{\left(\frac{\partial u}{\partial x}\right)^2 \frac{\partial^2 u}{\partial x^2} + \frac{\partial u}{\partial x}\frac{\partial u}{\partial y}\frac{\partial^2 u}{\partial x \partial y} + \left(\frac{\partial u}{\partial y}\right)^2 \frac{\partial^2 u}{\partial y^2} + \frac{\partial u}{\partial x}\frac{\partial u}{\partial y}\frac{\partial^2 u}{\partial x \partial y}}{\sqrt{\left(\frac{\partial u}{\partial x}\right)^2 + \left(\frac{\partial u}{\partial y}\right)^2}}$$

(3.60)

Next, we perform some approximations on (3.60). We consider that the second order derivatives do not vary too much [48], so the next approximation can be performed:

$$\nabla(\xi_u(\|\nabla u\|)) \cdot \nabla u \approx \xi_u'(\|\nabla u\|) \frac{\frac{\partial^2 u}{\partial x \partial y}\left(\frac{\partial u}{\partial x} + \frac{\partial u}{\partial y}\right)^2}{\sqrt{\left(\frac{\partial u}{\partial x}\right)^2 + \left(\frac{\partial u}{\partial y}\right)^2}} \approx \xi_u'\left(\sqrt{u_x^2 + u_y^2}\right) u_{xy}(u_x + u_x)$$

(3.61)

Then, the term $\xi_u'\left(\sqrt{u_x^2 + u_y^2}\right) u_{xy}(u_x + u_y)$ gets the following discretization:

$$\xi_u'\left(\sqrt{\frac{\left(u_{i+h,j}^n - u_{i-h,j}^n\right)^2}{4h^2} + \frac{\left(u_{i,j+h}^n - u_{i,j-h}^n\right)^2}{4h^2}}\right)$$

$$\frac{\left(u_{i+h,j+h}^n - u_{i+h,j-h}^n - u_{i-h,j+h}^n + u_{i-h,j-h}^n\right)\left(u_{i+h,j}^n - u_{i-h,j}^n + u_{i,j+h}^n - u_{i,j-h}^n\right)}{8h^3}$$

Next, by using this discretization of $div(\xi_u(\|\nabla u\|)\nabla u)$, the following implicit numerical approximation scheme is obtained for (3.54):

$$\frac{u_{i,j}^{n+\Delta t} - u_{i,j}^n}{\Delta t} = \lambda \xi_u\left(\sqrt{\frac{\left(u_{i+h,j}^n - u_{i-h,j}^n\right)^2}{4h^2} + \frac{\left(u_{i,j+h}^n - u_{i,j-h}^n\right)^2}{4h^2}}\right)$$

$$\frac{u_{i+h,j}^n + u_{i-h,j}^n + u_{i,j+h}^n + u_{i,j-h}^n - 4u_{i,j}^n}{h^2}$$

$$+ \lambda \xi_u'\left(\sqrt{\frac{\left(u_{i+h,j}^n - u_{i-h,j}^n\right)^2}{4h^2} + \frac{\left(u_{i,j+h}^n - u_{i,j-h}^n\right)^2}{4h^2}}\right)$$

$$\frac{\left(u_{i+h,j+h}^n - u_{i+h,j-h}^n - u_{i-h,j+h}^n + u_{i-h,j-h}^n\right)\left(u_{i+h,j}^n - u_{i-h,j}^n + u_{i,j+h}^n - u_{i,j-h}^n\right)}{8h^3}$$

$$- \rho\left(u_{i,j}^n - u_{i,j}^0\right)$$

(3.62)

One may consider $h = 1$ and $\Delta t = 1$. Next, the implicit discretization (3.62) is transformed into the following iterative explicit numerical approximation scheme:

$$u_{i,j}^{n+1} = u_{i,j}^n (1 - \rho) + \lambda \xi_u \left(\frac{\sqrt{\left(u_{i+1,j}^n - u_{i-1,j}^n\right)^2 + \left(u_{i,j+1}^n - u_{i,j-1}^n\right)^2}}{2} \right)$$

$$+ \left(u_{i+1,j}^n + u_{i-1,j}^n + u_{i,j+1}^n + u_{i,j-1}^n - 4u_{i,j}^n\right)$$

$$+ \lambda \xi_u' \left(\frac{\sqrt{\left(u_{i+1,j}^n - u_{i-1,j}^n\right)^2 + \left(u_{i,j+1}^n - u_{i,j-1}^n\right)^2}}{2} \right)$$

$$\frac{\left(u_{i+1,j+1}^n - u_{i+1,j-1}^n - u_{i-1,j+1}^n + u_{i-1,j-1}^n\right)\left(u_{i+1,j}^n - u_{i-1,j}^n + u_{i,j+1}^n - u_{i,j-1}^n\right)}{8} + \rho u_{i,j}^0$$

(3.63)

for each $i \in \{1, \ldots, I\}$, $j \in \{1, \ldots, J\}$ and $n \in \{0, \ldots, N\}$ [48].

The finite difference-based explicit numerical approximation algorithm (3.63) is consistent to the anisotropic diffusion model (3.54) and is converging fast to its solution representing the restored image. It has been used in our successful denoising experiments, being applied on hundreds digital images affected by additive Gaussian noise [48].

Equation (3.47) determines a class of variational schemes, each one depending on a parameter selection. The parameter values assuring an optimal filtering, $\lambda = 1.2$, $\rho = 0.3$, $\eta = 0.2$, $\beta = 0.7$, $\alpha = 1.3$, $\delta = 4$, $\zeta = 0.5$, $N = 12$, are empirically detected.

The restoration tests prove that our PDE-based filter reduces the additive noise considerably, while preserving the edges and other details. Also, it avoids or alleviates the unintended effects, such as image blurring, blocky effect and the speckle noise, and operates quite fast, having a execution time of less than 1 s [48].

Method comparison have been also performed. The performance of the proposed technique has been measured using well-known similarity metrics, such as PSNR, MSE and SSIM. The average PSNR values provided by our approach and other smoothing methods are displayed in Table 3.4. The proposed variational denoising framework outperforms many existing PDE-based and conventional restoration schemes, achieving higher PSNR values, as one can see in the next table.

Some image denoising results obtained by these restoration schemes on the $[512 \times 512]$ *Elaine* image are displayed in Fig. 3.5. The original is depicted in (a) while its version affected by an amount of Gaussian noise given by $\mu = 0.04$ and *variance* $= 0.05$, is displayed in (b). The output produced by the $[3 \times 3]$ two-dimension conventional filters are displayed in (c) and (d). The filtering results produced by the PDE and variational models are displayed in (e)—Perona-Malik 1, (f)—Perona-Malik 2, (g)—TV Denoising and (h)—our variational denoising approach providing the best image enhancement.

Table 3.4 Method
comparison: average PSNR
values

Restoration model	Average PSNR value (dB)
The proposed variational method	27.33
Average filter	25.63
2D Gaussian filter	25.47
Perona-Malik 1	26.85
Perona-Malik 2	26.85
TV Denoising (ROF)	27.14

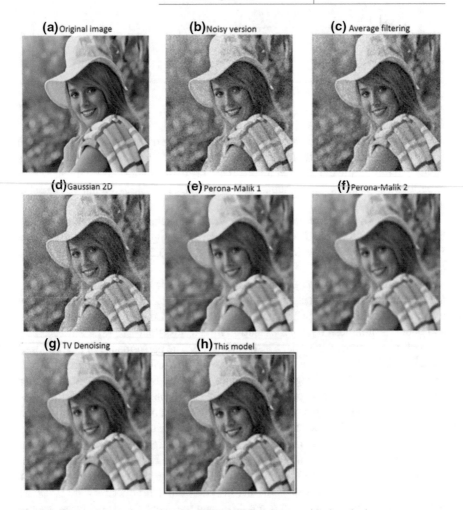

Fig. 3.5 Restoration results produced by several PDE and conventional methods

Other variational denoising schemes producing nonlinear second-order diffusion models have also been developed by us and disseminated in [49, 50]. However, they are outperformed by the variational framework described here.

3.2.3 Nonlinear Second-Order Hyperbolic PDE-Based Restoration Schemes

In the Sects. 2.2 and 2.3 there are described some of our linear hyperbolic second-order PDE models for image denoising and restoration. We have mentioned in those sections that some nonlinear, and more effective, versions of those hyperbolic schemes can be developed. We have constructed these nonlinear hyperbolic models and disseminated them in several past works [51, 52].

So, in [51] we propose such a nonlinear PDE-based restoration model for image restoration, which is composed on a second-order hyperbolic diffusion equation and boundary conditions. The considered PDE-based model has the following form:

$$
\begin{cases}
\alpha \frac{\partial^2 u}{\partial t^2} + \beta^2 \frac{\partial u}{\partial t} - div\left(\psi_{K(u)}(\|\nabla u\|)\nabla u\right) + \lambda(u - u_0) = 0 \\
u(0, x, y) = u_0(x, y) \\
\frac{\partial u}{\partial t}(0, x, y) = u_1(x, y) \\
u(t, x, y) = 0, \forall t \geq 0, (x, y) \in \partial \Omega
\end{cases}
\quad , \quad (x, y) \in \Omega \quad (3.64)
$$

where the coefficients $\alpha, \beta \in (0, 1]$ and $\lambda \in (0, 0.4]$, and u_0 represents the observed image.

The edge-stopping function $\psi_{K(u)} : [0, \infty] \to [0, \infty]$ is modeled as follows:

$$
\psi_{K(u)}(s) =
\begin{cases}
\xi \sqrt{\frac{K(u)}{\gamma s^2 + \eta}}, & \text{if } s > 0 \\
1, & \text{if } s = 0
\end{cases}
\quad (3.65)
$$

where $\xi, \gamma \in (0, 1]$, $\eta \in (0, 6)$ and the diffusivity conductance is constructed, as following:

$$
K(u) = \zeta |median(\|\nabla u\|) - \varepsilon| + v\, ord(u) \quad (3.66)
$$

where $\zeta, \varepsilon \in (0.2, 3]$, $v \in (0, 0.6)$, $ord(u)$ returns the order (time) of the current state of u in the evolving image sequence.

The function $\psi_{K(u)}$ is properly chosen, since it satisfies the main requirements of a proper diffusion-based denoising process. Obviously, this diffusivity function is always positive, since $\forall s > 0, \xi \sqrt{\frac{K(u)}{\gamma s^2 + \eta}} > 0$. It is also monotonically decreasing, because for $\forall s_1 \geq s_2$, we have: $\psi_{K(u)}(s_1) = \xi\sqrt{\frac{K(u)}{\gamma s_1^2 + \eta}} \leq \xi\sqrt{\frac{K(u)}{\gamma s_1^2 + \eta}} = \psi_{K(u)}(s_2)$. The convergence to zero is also satisfied, since $\lim\limits_{s \to \infty} \psi_{K(u)}(s) = 0$ [51].

The hyperbolic denoising model (3.64) removes successfully the diffusion effect in the vicinity of the image boundaries and corners. Thus, it provides sharper edges than the parabolic PDE-based filters.

The third component of this restoration scheme, $\lambda(u - u_0)$, has the role to stabilize the optimal filtering output, preventing the further deterioration of the smoothed image. This PDE model (3.64) has also the localization property, which means its solution is propagating with finite speed. The existence and uniqueness of this solution, corresponding to the recovered image, have also been rigorously investigated [51].

The well-posedness of the nonlinear second-order diffusion model is treated by performing an integration operation on the hyperbolic equation. Thus (3.64) can be re-written as an integral equation, as follows:

$$
\begin{cases}
\frac{\partial u}{\partial t}(t, x, y) - \int\limits_0^t e^{\frac{\beta^2}{\alpha}(t-s)} \begin{pmatrix} div\big(\psi_{K(u)}(\nabla u(s, x, y))\nabla u(s, x, y)\big) + \\ \lambda(u(s, x, y) - u_0(x, y)) \end{pmatrix} ds = u_1(x, y) \\
u(0, x, y) = u_0(x, y) \\
u(t, x, y) = 0, \text{on}(0, T) \times \partial\Omega
\end{cases}
$$

$$(3.67)$$

This equivalent problem is ill-posed, in general, but we consider a modified version of it that can become well-posed under some certain assumptions. So, (3.67) is transformed into the following model:

$$
\begin{cases}
\frac{\partial u}{\partial t} - \varepsilon\Delta u - \int\limits_0^t \big(div\big(\psi_{K(u)}(\|\nabla u\|)\nabla u\big) + \lambda(u - u_0)\big)ds = 0 \\
u(0, x, y) = u_0(x, y) \\
u(t, x, y) = 0, \text{on}(0, T) \times \partial\Omega
\end{cases}
$$

$$(3.68)$$

The new integral scheme has a solution for $\varepsilon > 0$ if the next conditions hold [53]:

$$
\begin{cases}
\big(\psi_{K(u)}(\|v_1\|)v_1 - \psi_{K(u)}(\|v_2\|)v_2\big)(v_1 - v_2) \geq 0, \quad \forall v_1, v_2 \in R^2 \\
\exists C, c : C \geq \psi_{K(u)}(r) \geq c > 0, \quad \forall r \geq 0 \\
\psi_{K(u)} - continuous
\end{cases}
$$

$$(3.69)$$

First condition means that function $u \to \psi_{K(u)}(u)$ u is monotone in R^2, so it reduces to:

$$
\frac{\partial\big(\psi_{K(u)}(r)r\big)}{\partial r} = \psi'_{K(u)}(r)r + \psi_{K(u)}(r) \geq 0, \quad \forall r \in R^+
$$

$$(3.70)$$

If $r = 0$, then $\psi_{K(u)}(r) = 1$, so (3.70) is satisfied. Otherwise, if $r > 0$, then $\psi_{K(u)}(r) = \xi \sqrt{\frac{K(u)}{\gamma r^2 + \eta}}$, therefore $\psi'_{K(u)}(r) r + \psi_{K(u)}(r) = \frac{\xi \gamma K(u)}{(\gamma r + \eta)^2 \sqrt{\frac{K(u)}{\gamma r + \eta}}} r + \xi \sqrt{\frac{K(u)}{\gamma r^2 + \eta}}$, so

$\frac{\partial}{\partial r}\left(\psi_{K(u)}(r) r\right) \geq 0, \forall r \geq 0$, so the condition (3.70) holds in this case too.

Second condition of (3.69) requires the function $\psi_{K(u)}$ to be bounded. The values c and C exists, since $\xi \sqrt{\frac{K(u)}{\eta}} > \psi_{K(u)}(s) > 0, \forall s > 0$. The third condition also holds, because $\psi_{K(u)}$ is a continuous function [47].

Under these assumptions the integral problem (3.68) has a unique weak solution u^* in sense of distributions [53], which means:

$$u^* \in L^\infty\left(0, T; H_0^1(\Omega)\right), \frac{\partial u}{\partial t} \in L^2\left(0, T; L^2(\Omega)\right) \tag{3.71}$$

and

$$div\left(\psi_{K(u)}(\|\nabla u\|)\nabla u\right) \in L^\infty\left(0, T; L^2(\Omega)\right) \tag{3.72}$$

Although the integral model (3.68) is somewhat different than (3.67), it could be considered as a good approximation of it. Also, the solution of (3.68) converges in a certain weak sense to a solution of (3.67) for $\varepsilon \to 0$.

A finite difference based numerical approximation scheme which converges fast to that solution is then proposed [47]. The hyperbolic equation can be re-written as:

$$\alpha \frac{\partial^2 u}{\partial t^2} + \beta^2 \frac{\partial u}{\partial t} - \psi_{K(u)}(\|\nabla u\|)\Delta u - \nabla\left(\psi_{K(u)}(\|\nabla u\|)\right) \cdot \nabla u + \lambda(u - u_0) = 0 \tag{3.73}$$

Its first component, $\alpha \frac{\partial^2 u}{\partial t^2} + \beta^2 \frac{\partial u}{\partial t}$, is discretized as following:

$$D_1 = \alpha \frac{u^{n+\Delta t}(i, j) + u^{n-\Delta t}(i, j) - 2u^n(i, j)}{\Delta t^2} + \beta^2 \frac{u^{n+\Delta t}(i, j) - u^{n-\Delta t}(i, j)}{2\Delta t}. \tag{3.74}$$

that leads to

$$D_1 = \frac{(2\alpha + \beta^2 \Delta t)}{2\Delta t^2} u^{n+\Delta t}(i, j) + \frac{(2\alpha - \beta^2 \Delta t)}{2\Delta t} u^{n-\Delta t}(i, j) - \frac{2u^n(i, j)}{\Delta t^2} \tag{3.75}$$

The second component, $\psi_{K(u)}(\|\nabla u\|)\Delta u = \psi_{K(u)}(\|\nabla u\|)\left(\frac{\partial^2 u}{\partial x^2} + \frac{\partial^2 u}{\partial y^2}\right)$, is approximated as:

$$D_2 = \psi_{K(u)}\left(\|\nabla u^n\|\right) \frac{u^n(i + h, j) + u^n(i - h, j) + u^n(i, j + h) + u^n(i, j - h) - 4u^n(i, j)}{h^2} \tag{3.76}$$

and for the third one we use a discretization process similar to that in (3.59)–(3.61), and obtain:

$$D_3 = \frac{\partial \psi_{K(u)}}{\partial s}\left(\|\nabla u^n\|\right)\frac{u^n(i+h, j+h) - u^n(i+h, j-h) - u^n(i-h, j+h) + u^n(i-h, j-h)}{h^2}$$
$$\cdot \frac{u^n(i+h, j) - u^n(i-h, j) + u^n(i, j+h) - u^n(i, j-h)}{2h} \tag{3.77}$$

Thus, we get the next numerical approximation for Eq. (3.73): $D_1 - D_2 - D_3 + \lambda(u^n(i, j) - u_0(i, j)) = 0$. For $h = 1$ and $\Delta t = 1$, it leads to the next iterative explicit numerical approximation scheme:

$$u^{n+1}(i, j) = \frac{4}{2\alpha + \beta^2}u^n(i, j) - \frac{2\alpha - \beta^2}{2\alpha + \beta^2}u^{n-1}(i, j)\psi_{K(u)}\left(\nabla u^n\right)$$
$$\left(u^n(i+1, j) + u^n(i-1, j) + u^n(i, j+1) + u^n(i, j-1) - 4u^n(i, j)\right)$$
$$+ \frac{1}{8}\psi'_{K(u)}\left(\nabla u^n\right)$$
$$\left(u^n(i+1, j+1) - u^n(i+1, j-1) - u^n(i-1, j+1) + u^n(i-1, j-1)\right)$$
$$\left(u^n(i+1, j) - u^n(i-1, j) + u^n(i, j+1) - u^n(i, j-1)\right)$$
$$- \lambda\left(u^n(i, j) - u_0(i, j)\right), \forall n \geq 1 \tag{3.78}$$

This iterative denoising scheme is consistent to the PDE hyperbolic model, stable and fast converging. It applies repeatedly the process (3.78), for $i = 0, 1, \ldots, N$, the number of iterations being quite low.

The proposed nonlinear hyperbolic diffusion-based filtering technique has been tested on hundreds of images corrupted with various amounts of additive noise. It removes effectively and fast the Gaussian noise, while preserving successfully the essential image details and overcoming the undesired effects. It avoids completely the blurring effect and alleviates the staircase effect. The optimal restoration results have been obtained by using the next model's parameters: $\alpha = 0.65$, $\beta = 0.5$, $\lambda = 0.2$, $\xi = 0.75$, $\gamma = 0.4$, $\eta = 5$, $\zeta = 0.25$, $\varepsilon = 1.5$, $\nu = 0.2$, $N = 15$.

Our second-order hyperbolic PDE-based method outperforms conventional filters, such as 2D Gaussian, Average or Wiener, by providing a better noise removal, overcoming the blurring effect and preserving the image features. It also outperforms our linear hyperbolic PDE-based denoising approach described in Sect. 2.2, representing an improved non-linearized version of it.

The parabolic second-order diffusion-based denoising techniques, such as Perona-Malik model, TV-ROF Denoising and the diffusion schemes derived from them, are also outperformed by this hyperbolic framework. As we have already mentioned, it preserves and defines the image edges much better than parabolic diffusion models. It also executes faster than them and reduces more the staircasing. The proposed restoration algorithm performs better than some nonlinear fourth-order PDE models, such as the You-Kaveh scheme, since it provides a much more effective deblurring and avoids the multiplicative noise that affects the filters of that type. As one can

Table 3.5 Method comparison: average PSNR values

Denoising technique	Average PSNR value (dB)
Nonlinear hyperbolic model	27.93
Linear hyperbolic PDE model	26.14
Perona-Malik scheme	25.28
TV Denoising	22.17
You and Kaveh model	26.88
Gaussian filter	21.97
Median 2D	24.32

see in Table 3.5, the hyperbolic PDE-based technique described here achieves higher PSNR values than PDE and non-PDE image filters.

An example of an image restored by our technique and compared to the denoising results of the methods specified in the table is described in Fig. 3.6. One can see that the proposed nonlinear hyperbolic PDE approach provides the best image enhancement.

Another second-order nonlinear hyperbolic diffusion model for image restoration, developed by us, is disseminated in [52]. It also represents an improved nonlinear version of our linear hyperbolic PDE model described in Sect. 2.2, being closely related to Eq. (2.17). The proposed model is composed of a second-order hyperbolic PDE and several boundary conditions:

$$\begin{cases} \gamma \frac{\partial^2 u}{\partial t^2} + \eta^2 \frac{\partial u}{\partial t} - \xi_u(\|\nabla u\|)\Delta u + \alpha(u - u_0) = 0 \\ u(0, x, y) = u_0(x, y) \\ u_t(0, x, y) = u_1(x, y) \\ u(t, x, y) = 0, \forall t \geq 0, (x, y) \in \partial\Omega \end{cases} \quad , \quad (x, y) \in \Omega \qquad (3.79)$$

where $\gamma, \eta \in (0, 1], \alpha \in (0, 0.4]$ and $\Omega \subset R^2$. The edge-stopping function ξ_u is modeled as:

$$\xi_u : [0, \infty] \to [0, \infty], \xi_u(s) = \frac{\beta}{\left(\frac{k(u)}{s}\right)^2 + k(u)\left|\ln\left(\frac{k(u)}{s}\right)\right|} \qquad (3.80)$$

where $\beta \in (1, 17]$, the conductance parameter function has the form $k(u(x, y, t)) = \varepsilon\mu(\|\nabla u\|) - vt$, where the coefficients $\varepsilon \in (2, 3]$ and $v \in (0, 1)$.

The diffusivity function is properly constructed, being positive, monotonically decreasing and convergent to zero [52]. A mathematical treatment of this second-order PDE-based framework is also provided, the well-posedness of our hyperbolic model being demonstrated [52]. So, the PDE in (3.79) is equivalent to the following equation:

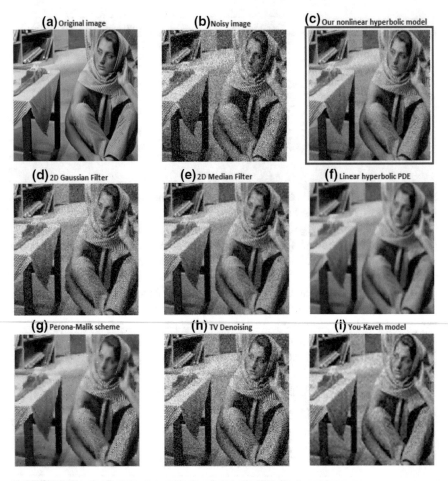

Fig. 3.6 Restoration results produced by several methods on *Barbara* image

$$\gamma \frac{\partial^2 u}{\partial t^2} + \eta^2 \frac{\partial u}{\partial t} - div\big(\widetilde{\xi}(u)\big(\|\nabla u^2\|\big)\big) + \alpha(u - u_0) = 0 \qquad (3.81)$$

where $\widetilde{\xi}'_u(s) = \xi_u(\sqrt{s})$, $\forall s \geq 0$. Equation (3.81) accepts solutions if some certain conditions are met. Thus, we must have $\widetilde{\xi}'_u(s) \geq 0$ that leads to $\xi_u(s) \geq 0$, that is a satisfied condition. Also, ξ_u must satisfy the bounding condition:

$$\exists K > 0 : \xi_u(s) \leq K\big(s^2 + s + 1\big), \quad \forall s \geq 0 \qquad (3.82)$$

If this condition is also satisfied, then there exists a solution to the model (3.79) in some generalized sense [54]. The relation (3.82) is equivalent to $\xi_u(s) = \dfrac{\beta}{\left(\frac{k(u)}{s}\right)^2 + k(u)\left|\ln\left(\frac{k(u)}{s}\right)\right|} \leq K\big(s^2 + s + 1\big)$, which leads to $K \geq \dfrac{\beta}{(s^2+s+1)\left(\left(\frac{k(u)}{s}\right)^2 + k(u)\left|\ln\left(\frac{k(u)}{s}\right)\right|\right)}$.

Obviously, such a K value exists for any $s > 0$, so (3.82) holds. In [53, 54] it is proved the existence and uniqueness of a solution $u = u(t, x)$, such that:

$$u \in L^2\left(0, T; H_0^1(\Omega)\right), \frac{\partial u}{\partial t} \in L^2((0, T) \times \Omega) \tag{3.83}$$

where H_0^1 is the standard Sobolev space [53].

Therefore, the considered nonlinear hyperbolic PDE-based model is well-posed and it has also the localization property, its unique and weak solution propagating with finite speed. Because this localization property, the evolving image will remain quite close to the observed one.

A finite difference based numerical approximation scheme is constructed to compute this weak solution. We use the same grid and quantization as in the previous cases. The hyperbolic equation in (3.79) is discretized as:

$$\gamma \frac{u^{n+\Delta t}(i, j) + u^{n-\Delta t}(i, j) - 2u^n(i, j)}{\Delta t^2} + \eta^2 \frac{u^{n+\Delta t}(i, j) - u^{n-\Delta t}(i, j)}{2\Delta t}$$

$$- \xi_u \left(\sqrt{\left(\frac{u^n(i+h, j) - u^n(i-h, j)}{2h}\right)^2 + \left(\frac{u^n(i, j+h) - u^n(i, j-h)}{2h}\right)^2} \right)$$

$$\frac{u^n(i+h, j) + u^n(i-h, j) + u^n(i, j+h) + u^n(i, j-h) - 4u^n(i, j)}{h^2}$$

$$+ \alpha\left(u^n(i, j) - u^0(i, j)\right) = 0 \tag{3.84}$$

If we take $h = 1$ and $\Delta t = 1$, (3.84) leads to the next explicit numerical approximation scheme:

$$u^{n+1}(i, j) = \frac{2\alpha - 4\gamma}{2\gamma + \eta^2} u^n(i, j) + \frac{\eta^2 - 2\gamma}{2\gamma + \eta^2} u^{n-1}(i, j)$$

$$- \frac{2}{2\gamma + \eta^2} \xi_u \left(\sqrt{(u^n(i+1, j) - u^n(i-1, j))^2 + (u^n(i, j+1) - u^n(i, j-1))^2} \right)$$

$$\left(u^n(i+1, j) + u^n(i-1, j) + u^n(i, j+1) + u^n(i, j-1) - 4u^n(i, j)\right)$$

$$- \frac{2\alpha}{2\gamma + \eta^2} u^0(i, j) \tag{3.85}$$

for $i \in \{1, \ldots, I\}$, $j \in \{1, \ldots, J\}$ and $n \in \{0, \ldots, N\}$. The iterative numerical approximation algorithm (3.85) is consistent to the hyperbolic PDE model and stable, so it converges fast to the model's solution [52].

The proposed nonlinear second-order PDE-based denoising technique has been tested on many images corrupted with various amounts of Gaussian noise, such as those of the Volume 3 of the USC-SIPI database. It not only removes successfully the Gaussian noise, but also preserves the image details and provides sharper edges. It also overcomes the blurring effect and alleviates the staircasing. The following set of parameter values provides optimal image denoising results:

$$\gamma = 2.3, \eta = 1.5 \beta = 1.2, \alpha = 0.12, \varepsilon = 0.8 v = 0.2, N = 19$$

Table 3.6 Method comparison: average SSIM values

Filtering approach	Average SSIM value
This technique	0.6342
Average filter	0.5625
Gaussian filter	0.5498
LLMMSE filter	0.6239
Perona-Malik scheme	0.6183
Total Variation Denoising	0.5857
You-Kaveh model	0.6014

The performance of our approach has been assessed by using measures like Structural Similarity Image Metric and Peak Signal-to-Noise Ratio [55]. The proposed hyperbolic diffusion-based method outperforms both the conventional and PDE-based restoration techniques, achieving higher average values for the performance measures. One can see some method comparison results in Table 3.6, which contains the average SSIM values produced by our model and other PDE and non-PDE denoising algorithms, and in Fig. 3.7, which displays the restoration results obtained by these filtering approaches on the *Lenna* image corrupted by an amount of additive noise.

3.3   Nonlinear Fourth-Order PDE Models for Image Restoration

Numerous nonlinear fourth-order PDE-based image restoration models have been developed by us in the last years. Our main contributions in this domain will be presented in this section. These contributions, representing novel fourth-order diffusion-based denoising techniques have been disseminated in papers published in recognized international journals or volumes of international conferences.

The mathematical investigations that treat the well-posedness of these nonlinear fourth-order PDE models, by analyzing the existence and uniqueness of the (weak) solutions, are also described in this section. Consistent and stable finite-difference based numerical approximation schemes are constructed for this models and the results of our numerical experiments and method comparison are also addressed here.

The first subsection describes some parabolic fourth-order PDE-based restoration models and variational schemes that lead to these models, developed by us. Our nonlinear hyperbolic fourth-order diffusion-based smoothing techniques approaches will be presented in the second subsection. The last subsection describes some nonlinear hybrid PDE-based approaches involving fourth-order diffusions.

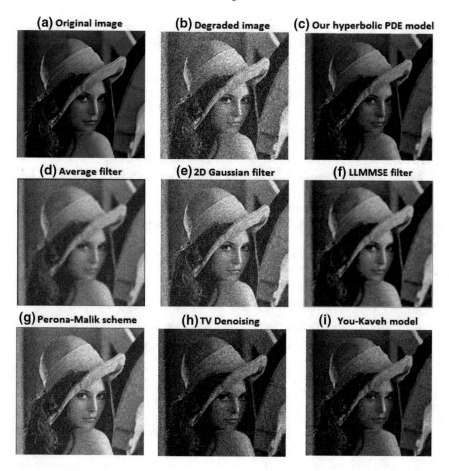

Fig. 3.7 Method comparison: denoising results achieved by several models

3.3.1 Variational and Parabolic PDE-Based Denoising Models

We have developed several nonlinear fourth-order parabolic PDE-based restoration approaches in our past works [56, 57]. In [56] such a denoising model, composed of a nonlinear fourth-order PDE and several boundary conditions, is proposed under the following form:

$$
\begin{cases}
\frac{\partial u}{\partial t} + \nabla^2(\psi^u(\|\nabla u\|)\Delta u) + \varepsilon(u - u_0) = 0 \\
u(0, x, y) = u_0(x, y) \\
u(t, x, y) = 0, \forall t \geq 0, (x, y) \in \partial\Omega \\
\Delta u(t, x, y) = 0, \forall(x, y) \in \partial\Omega
\end{cases}
, \quad (x, y) \in \Omega \qquad (3.86)
$$

where $\varepsilon \in (0, 1)$, $\Omega \subseteq R^2$ and u_0 is the observed image, affected by Gaussian noise. The diffusivity function, $\psi^u : [0, \infty] \rightarrow [0, \infty]$, is properly chosen for the diffusion process, as follows:

$$\psi^u(s) = \gamma \sqrt{\frac{K(u)}{\lambda \ln\left(s^2 + K(u)\right)^3 + \alpha}} \tag{3.87}$$

where $K(u) = |\xi\mu(\|\nabla u\|) - vt|$ and the coefficients $\alpha, \lambda \in [1, 3)$, $\xi, v \in (0, 1)$ [56].

So, the restored image is determined by solving the parabolic PDE given by (3.86) and (3.87). The existence and uniqueness of a solution of this nonlinear fourth-order PDE has been also investigated. In its current form, the considered PDE model is ill-posed, but we have proved it may accept solutions under some certain conditions [56].

Thus, one might define a generalized solution u for the model [56]. So, let us consider the sequence $\{u_n\}$ that is iteratively defined as:

$$\begin{cases} \frac{\partial u_n}{\partial t} + \Delta(\psi^u(\|\nabla u_{n-1}\|)\Delta u_n) + \varepsilon(u - u_0) = 0 \\ u_n(0, x, y) = u_0(x, y), \forall(x, y) \in \Omega \\ u_n(t, x, y) = \Delta u_n(t, x, y), \forall(x, y) \in \partial\Omega \end{cases} \tag{3.88}$$

Because ψ^u represents a bounded function, (3.88) has a unique solution u_n for each u, with the properties:

$$\frac{\partial u_n}{\partial t} \in L^2((0, T) \times \Omega); \Delta u_n \in L^2((0, T) \times \Omega) \tag{3.89}$$

The problem (3.88) represents the dynamic version of the next variational scheme:

$$\min_{u \in H^2(\Omega)} \left\{ \int_\Omega \psi^u(\|\nabla u_{n-1}\|)|\Delta u|^2 dxdy + \varepsilon \int_\Omega (u - u_0)^2 dxdy \right\} \tag{3.90}$$

By (3.88) we get the next estimate for u_n:

$$\frac{1}{2} \int_\Omega |u_n(t, x, y)|^2 dxdy + \int_0^t \int_\Omega \psi^u(\nabla u_{n-1})|\Delta u_n(r, x, y)|^2 dxdydr$$

$$+ \varepsilon \int_\Omega |u_n(t, x, y) - u_0(x, y)|^2 dxdy = \frac{1}{2} \int_\Omega |u_0(x, y)|^2 dxdy \tag{3.91}$$

It results that $\{u_n\}$ is bounded in $C\left([0, T]; L^2(\Omega)\right) \cap L^2\left(0, T; H^2(\Omega)\right)$ and $\left\{\frac{\partial u_n}{\partial t}\right\}$ is bounded in $L^2\left(0, T; H^{-1}(\Omega)\right)$ [47]. It follows that u_n is convergent in topology of $L^2\left(0, T; H^{-1}(\Omega)\right)$ to u as $n \to \infty$ [53, 58]. Hence

$$\psi^u(\|\nabla u_{n-1}\|) \to \psi^u(\|\nabla u\|), \text{ on } (0, T) \times \Omega \tag{3.92}$$

One can see that u satisfies the fourth-order PDE-based model in the weak sense, that is:

$$\begin{cases} \int\limits_\Omega \frac{\partial u}{\partial t}(t, x)\varphi(x)dx + \int\limits_\Omega \psi^u(\nabla u(t, x)\Delta u(t, x))\Delta\varphi(x)dx \\ \quad + \varepsilon \int\limits_\Omega (u(t, x, y) - u_0(x, y))\varphi(x)dx = 0, \quad \forall t \in (0, T) \\ u(0, x) = u_0(x) \end{cases} \tag{3.93}$$

for all $\varphi \in H^{-1}(\Omega)$ such that $\varphi = \Delta\psi^u = 0$ on Ω. Therefore, we may infer that under our assumptions on φ, the model (3.86) has a weak solution u in sense of (3.93) that can be computed as limit of $\{u_n\}$ iteratively defined by (3.88) [53, 58].

The nonlinear diffusion scheme is solved numerically by using a consistent finite difference-based discretization algorithm. It uses the same grid of size h and time step Δt, and the same quantization of space and time coordinates, as in the previous cases. We get the next explicit numerical approximation scheme:

$$u_{i,j}^{n+1} = u_{i,j}^n - \eta_{i+1,j}^n - \eta_{i-1,j}^n - \eta_{i,j+1}^n - \eta_{i,j-1}^n + 4\eta_{i,j}^n - \varepsilon\left(u_{i,j}^n - u_{i,j}^o\right) \tag{3.94}$$

where $\eta_{i,j}^n = \psi^u\left(\left|\nabla u_{i,j}^n\right|\right)\Delta u_{i,j}^n$ and the discrete Laplacian is computed as follows:

$$\Delta u_{i,j}^n = u_{i+1,j}^n + u_{i-1,j}^n + u_{i,j+1}^n + u_{i,j-1}^n - 4u_{i,j}^n \tag{3.95}$$

This fast-converging iterative scheme is then applied in our successful image restoration experiments with the following empirically detected parameter values: $\varepsilon = 0.2, \xi = 0.35, \nu = 0.14, \lambda = 2.7, \alpha = 4, N = 16$. The proposed approach removes the additive noise while preserving the edges and overcoming the staircasing and other unintended effects.

It outperforms the classic 2D filters, and also many state-of-the-art 2nd-order PDE denoising models, achieving better filtering results than Perona-Malik and TV Denoising alike schemes and avoiding the blocky effect that affects them. The proposed fourth-order diffusion method outperforms also the well-known fourth-order You-Kaveh model, by providing a better image deblurring and avoiding the speckle noise, while executing faster than it.

The average PSNR values, based on several hundreds of tested images, which are displayed in Table 3.7, illustrate the strength of the proposed technique. Some method comparison results are also depicted in Fig. 3.8. The denoising output produced by the

Table 3.7 Method comparison: average PSNRs

Image restoration scheme	PSNR (dB)
Our fourth-order PDE model	27.45
Average filter	25.23
2D Gaussian filter	25.14
Perona-Malik scheme	25.87
TV-ROF Denoising	26.07
You-Kaveh model	26.91

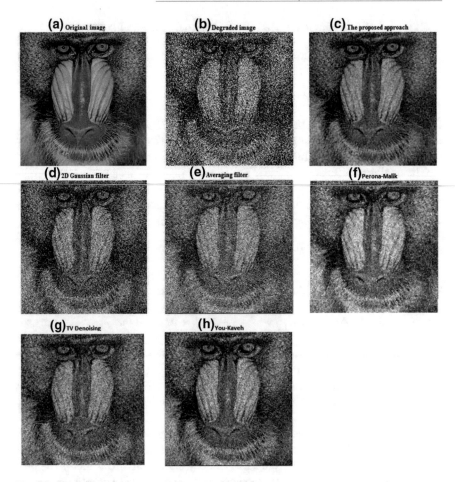

Fig. 3.8 The Baboon image restored by several models

proposed approach on the *Baboon* image corrupted by Gaussian noise and displayed in (c) looks better than the other restoration results.

Another effective nonlinear parabolic fourth-order diffusion based restoration framework is derived from a variational problem in [57]. There we consider the next variational model minimizing an energy cost functional:

$$u^* = \arg \min_u \int_\Omega \left(\frac{\alpha}{2} \xi_u(\|\Delta u\|) + \frac{\lambda}{2}(u - u_0)^2 \right) d\Omega \qquad (3.96)$$

where $\alpha, \lambda \in (0, 1)$ and $\Omega \subseteq R^2$. A fourth-order PDE model is derived from this variational scheme, by determining the next Euler-Lagrange equation:

$$\alpha \nabla^2 \left(\frac{\xi_u'(\|\Delta u\|)}{|\Delta u|} \nabla^2 u \right) + \lambda(u - u_0) = 0 \qquad (3.97)$$

We consider $\varphi_u(s) = \frac{1}{s} \frac{\partial \xi_u(s)}{\partial s}$ and obtains:

$$\alpha \nabla^2 \left(\varphi_u(|\Delta u|) \nabla^2 u \right) + \lambda(u - u_9) \qquad (3.98)$$

The gradient descent method is then applied and some boundary conditions are added, the next fourth-order PDE model being obtained:

$$\begin{cases} \frac{\partial u}{\partial t} + \alpha \Delta \left(\varphi_u(|\nabla^2 u|) \Delta u \right) + \lambda(u - u_0) = 0 \\ u(0, x, y) = u_0(x, y), \quad \forall (x, y) \in \Omega \\ u(t, x, y) = 0, \quad \forall (x, y) \in \partial \Omega \end{cases} \qquad (3.99)$$

where the diffusivity function $\varphi_u : [0, \infty) \to [0, \infty)$ is properly constructed as:

$$\varphi_u(s) = \frac{\beta}{\left(\frac{s}{v(u)} \right)^k + v(u) \left| \log \left(\frac{s}{v(u)} \right)^k \right|} \qquad (3.100)$$

where the conductance parameter $v(u) = \gamma(median(|\nabla u|) + \mu(\|\Delta u\|))$ and $\beta, k, \gamma \in [1, 3)$. The recovered image is determined by solving numerically this PDE-based problem. A finite difference-based discretization similar to that in the previous case produces the next explicit numerical approximation scheme:

$$u_{i,j}^{n+1} = (1 - \lambda)u_{i,j}^n - \alpha \left(\xi_{i+1,j}^n - \xi_{i-1,j}^n - \xi_{i,j+1}^n - \xi_{i,j-1}^n + 4\xi_{i,j}^n \right) - \lambda u_{i,j}^o \qquad (3.101)$$

The proposed variational technique reduces considerably the Gaussian noise and overcomes the staircasing and other unintended effects, preserving the edges and other essential details. It outperforms the conventional 2D filters, second-order PDE-based approaches like TV Denoising and P-M scheme, and also the You-Kaveh model. One can see some method comparison results ilustrating the performance of this technique in Table 3.8 and Fig. 3.9.

Table 3.8 Average PSNR values achieved by various methods

Denoising method	PSNR (dB)
This fourth-order PDE model	26.95
Average filter	25.43
Gaussian 2D filter	25.17
Median filter	25.53
Perona-Malik approach	25.91
TV Denoising	26.17
You-Kaveh scheme	26.73

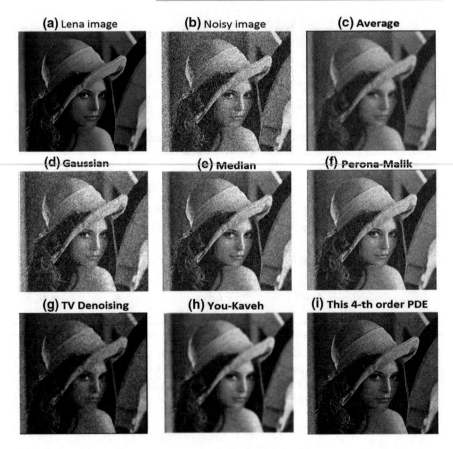

Fig. 3.9 *Lenna* image restored by our method and other approaches

3.3.2 Fourth-Order Hyperbolic PDE-Based Restoration Schemes

As previously mentioned in the last chapter, we have developed some nonlinear versions of the linear hyperbolic PDE denoising models described in Sects. 2.2 and 2.3, based not only on second-order PDEs, but also on fourth-order PDEs. These nonlinear fourth-order hyperbolic diffusion-based techniques are described in these subsection.

Thus, in [59] we propose a hyperbolic PDE-based image restoration approach that may be viewed as a nonlinear fourth-order version of the model (2.12) or as a fourth-order version of (2.17). The considered denoising model is composed of a hyperbolic PDE and some boundary conditions, as follows:

$$\begin{cases} \xi \frac{\partial^2 u}{\partial t^2} + \gamma^2 \frac{\partial u}{\partial t} + \Delta\big(\varphi_u(\|\nabla u\|)\nabla^2 u\big) + \lambda(u - u_0) = 0 \\ u(0, x, y) = u_0(x, y) \\ u_t(0, x, y) = u_1(x, y) \\ u(t, x, y) = 0, \forall t \geq 0, (x, y) \in \partial\Omega \end{cases} \quad , \quad (x, y) \in \Omega \quad (3.102)$$

where the parameters $\xi, \gamma, \lambda \in (0, 1]$, u_0 represents the observed image and u_1 represents a velocity modification of it [59]. The diffusivity function $\varphi_u : [0, \infty) \to [0, \infty)$ has the form:

$$\varphi_u(s) = \delta \sqrt{\frac{\psi(u)}{\alpha \log 10(s^2 + \psi(u))^k + \beta}} \quad (3.103)$$

where the conductance parameter is

$$\psi(u) = |\eta\mu(\|\nabla u\|) - \varepsilon| \quad (3.104)$$

the coefficients $\delta, \alpha, \eta \in (0, 1.5)$, $k, \beta, \varepsilon \in [2, 5)$. This function is properly constructed for an effective restoration. It is positive: $\varphi_u(s) > 0, \forall s \geq 0$. Also, it is a monotonically decreasing function, because we have $\forall s_1 \geq s_2, \varphi_u(s_1) = \delta \sqrt{\frac{\psi(u)}{\alpha \log 10(s_1^2 + \psi(u))^k + \beta}} \leq \varphi_u(s_2) = \delta \sqrt{\frac{\psi(u)}{\alpha \log 10(s_2^2 + \psi(u))^k + \beta}}$, and it converges to zero, since $\lim\limits_{s \to \infty} \varphi_u(s) = 0$. The function $\varphi_u(s)$ is also bounded, since $\exists b_1, b_2 \geq 0 : b_1 \leq \varphi_u(s) \leq b_2, \forall s \geq 0$. It also represents a Lipschitz function, since its derivative $\varphi_u'(s)$ is bounded. These conditions are important for the mathematical treatment of the PDE model, which analyze its well-posedness [59].

A rigorous mathematical investigation is performed in [59], where we demonstrate that the proposed fourth-order diffusion model admits a unique and weak solution under some certain assumptions. Also, our denoising scheme has the localization property, its solution propagating with finite speed [60]. That solution is determined by solving numerically the model (3.101) using a finite difference-based algorithm.

We consider the same space grid size of h and time step Δt, and the quantization $x = ih, y = jh, t = n\Delta t, i \in \{0, \ldots, I\}, j \in \{0, \ldots, J\}, n \in \{1, \ldots, N\}$, and perform the discretization of the partial differential equation in (3.102). The component $\xi \frac{\partial^2 u}{\partial t^2} + \gamma^2 \frac{\partial u}{\partial t} + \lambda(u - u_0) = 0$ is approximated first as $\xi \frac{u_{i,j}^{n+\Delta t} + u_{i,j}^{n-\Delta t} - 2u_{i,j}^n}{\Delta t^2} + \gamma^2 \frac{u_{i,j}^{n+\Delta t} - u_{i,j}^{n-\Delta t}}{2\Delta t} + \lambda\left(u_{i,j}^n - u_{i,j}^0\right)$. The second component is then discretized. One computes $\varphi_{i,j}^n = \varphi_u\left(\left\|\nabla u_{i,j}^n\right\|\right)\nabla^2 u_{i,j}^n$, with $\nabla^2 u_{i,j}^n = \Delta u_{i,j}^n = \frac{u_{i+h,j}^n + u_{i-h,j}^n + u_{i,j+h}^n + u_{i,j-h}^n - 4u_{i,j}^n}{h^2}$.

Then, one applies the Laplacian operator again and obtain the next discretization:

$$\nabla^2 \varphi_{i,j}^n = \Delta\left(\varphi_u\left(\left\|\nabla u_{i,j}^n\right\|\right)\nabla^2 u_{i,j}^n\right) = \frac{\varphi_{i+h,j}^n + \varphi_{i-h,j}^n + \varphi_{i,j+h}^n + \varphi_{i,j-h}^n - 4\varphi_{i,j}^n}{h^2}$$

$$(3.105)$$

We consider $h = \Delta t = 1$ and get the implicit numerical approximation scheme:

$$\xi\left(u_{i,j}^{n+1} + u_{i,j}^{n-1} - 2u_{i,j}^n\right) + \gamma^2 \frac{u_{i,j}^{n+1} - u_{i,j}^{n-1}}{2} + \Delta\varphi_{i,j}^n + \lambda\left(u_{i,j}^n - u_{i,j}^0\right) = 0 \quad (3.106)$$

It leads to the following explicit numerical approximation algorithm:

$$u_{i,j}^{n+1} = \frac{4\xi - 2\lambda}{2\xi + \gamma^2} u_{i,j}^n - \frac{2\xi - \gamma^2}{2\xi + \eta^2} u_{i,j}^{n-1} + \frac{2\Delta\varphi_{i,j}^n}{2\xi + \gamma^2} + \frac{2\lambda}{2\xi + \gamma^2} u_{i,j}^0, \quad \forall n \in [1, N]$$

$$(3.107)$$

This iterative discretization scheme starts with $u^1 = u^0 = u_0$ and applies repeatedly the procedure (3.107) on the evolving image, for each $n \in \{1, \ldots, N\}$. It is stable and consistent to the nonlinear hyperbolic PDE model (3.101) and converges quite fast to its solution.

We have successfully tested the proposed filtering technique on hundreds of images corrupted by Gaussian noise. Some important image collections, such as the Volume 3 of the USC-SIPI database have been used in our simulations. We have determined through empirical observation, the optimal set of parameters: $\delta = 0.5, \xi = 0.8, \gamma = 0.7, \lambda = 0.2, \alpha = 1.4, \eta = 0.3, k = 3, \beta = 4, \varepsilon = 3, N = 14$.

The performed experiments show that our hyperbolic PDE-based restoration method removes the addtive Gaussian noise and avoids the multiplicative noise, while preserving the edges and other features. It also overcomes the unintended effects of blurring and staircasing. The proposed approach executes fast (see the low N value), its running time being less than 1 s. Its performance has been assessed using similarity metrics such as PSNR, MSE and SSIM [55].

Our nonlinear fourth-order diffusion model outperforms the conventional filters, like Average, Gaussian 2D, Wiener and even more effective filters such as LLMM-SE–Lee, and the linear PDE-based methods, since it overcomes the image blurring

Table 3.9 Method comparison: average PSNR values

Restoration approach	Average PSNR values (dB)
Our fourth-order PDE model	28.05
Averaging filter	25.43
Gaussian 2D	25.27
2D Wiener filter	26.51
LLMMSE-Lee	27.63
Perona-Malik model	26.84
Total Variation—ROF	26.79
You-Kaveh scheme	27.31

and preserves the essential details. Also, unlike those schemes, the proposed model has the localization property.

Our fourth-order PDE-based approach outperforms also the state of the art second-order anisotropic diffusion schemes, such as the both versions of the Perona-Malik and the TV-ROF model, providing a more effective noise removal, running faster and overcoming the blocky effect. It also performs better than isotropic diffusion You-Kaveh model and the parabolic fourth-order denoising methods derived from it, because it avoids the blurring effect and the undesired speckle noise, and executes considerably faster than them.

The averaged PSNR values obtained by the proposed framework and other filtering methods are registered in Table 3.9. Our fourth-order PDE-based algorithm obtains higher PSNR values than both PDE and non-PDE models. The image restoration example described in Fig. 3.10 illustrates also its denoising performance.

Another hyperbolic PDE-based denoising scheme developed by us may be viewed as a nonlinear fourth-order version of the linear anisotropic diffusion model (2.18) described in Sect. 2.3. The restoration model that is disseminated in [61] has the following form:

$$
\begin{cases}
\alpha \frac{\partial^2 u}{\partial t^2} + \beta^2 \frac{\partial u}{\partial t} + \Delta(\psi_u(\|\nabla u\|)\Delta u) + J \cdot \Delta u = 0 \\
u(0, x, y) = u_0(x, y) \\
u_t(0, x, y) = u_1(x, y) \\
u(t, x, y) = 0, \forall t \geq 0, (x, y) \in \partial\Omega
\end{cases}
\quad , \quad (x, y) \in \Omega \qquad (3.108)
$$

where $\alpha, \beta \in (0, 3]$ and the 2D function J has the form:

$$
J(x, y) = \left(e^{-p(x+y)}, e^{-q(x^2+y^2)} \right) \qquad (3.109)
$$

with $p, q > 0$. The diffusivity function $\psi_u : [0, \infty) \to (0, \infty)$ is properly constructed as a positive and monotonically decreasing function having the following form:

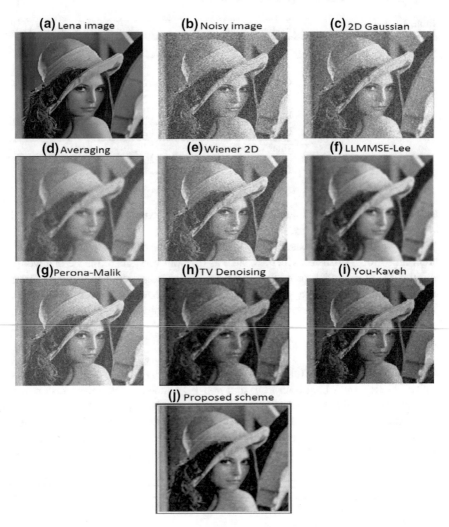

Fig. 3.10 *Lenna* denoising produced by our approach (j) and other filtering schemes

$$\psi_u(s) = \zeta \sqrt{\frac{\eta(u)}{\gamma \ \ln\big(s^k + \eta(u)\big)^t + \lambda}} \qquad (3.110)$$

where $\eta(u) = |\varphi\mu(\nabla u) - \delta|$, $t \in \{2, 3, 4\}$, $\lambda, \gamma, \zeta \in (1, 8]$ and $\varphi, \delta \in (0, 1)$.

This fourth-order PDE model is well-posed, admitting a unique and weak solution representing the recovered image [61]. The solution is computed by using an explicit fast-converging finite-difference based numerical approximation scheme which is constructed similarly to that in the previous case [61].

Table 3.10 Method comparison: average PSNR values

Denoising technique	PSNR (dB)
The proposed PDE model	27.16
Average filter	25.28
2D Gaussian	25.14
Perona-Malik scheme	25.97
TV Denoising	26.17
You-Kaveh model	26.91

This nonlinear hyperbolic PDE-based restoration approach has been successfully applied on numerous images affected by additive noise. It reduces considerably this type of noise and preserves all the image details, since it overcomes the unintended effects, like blurring, staircasing or speckling. The proposed algorithm outperforms many state-of-the-art nonlinear second-order PDE-based denoising models, since it does not generate the staircase effect, and also provides better results than the parabolic fourth-order PDE denoising models, like the You-Kaveh scheme, because it produces a better deblurring, avoids the multiplicative noise and also produces sharper boundaries. See the average PSNR values achieved by our approach and other PDE and non-PDE restoration models in Table 3.10.

A method comparison example is displayed in Fig. 3.11, which depicts the smoothing results provided by several filters on a noisy *Baboon* image.

3.3.3 Hybrid Fourth-Order Diffusion-Based Filtering Approaches

We have discussed in Sect. 3.1.2 about the hybrid restoration approaches involving second and fourth order diffusions. Since the individual PDE denoising models still have some drawbacks, like generating undesired effects, combining them to other PDE-based schemes or non-PDE filters may overcome these drawbacks and produce better restoration solutions.

Such a hybrid denoising solution considered by us is based on a combination of nonlinear second- and fourth-order diffusions. In [62] we propose a variational model that minimizes the following energy cost functional:

$$J(u) = \int_{\Omega} \left(\varphi_1(\|\nabla u\|) + \varphi_2(\|\nabla^2 u\|) + \frac{\lambda}{2}(u - u_0)^2 \right) d\Omega \qquad (3.111)$$

where φ_1 and φ_2 represent the two regularizers, $\lambda \in (0, 1)$, $\Omega \subseteq R^2$ and u_0 is the observed image. The first smoothing term of the energy functional J assures that the restoration is performed along the gradient direction, while the second one uses the Laplacian to approximate the observed image with a planar one. Then, a nonlinear

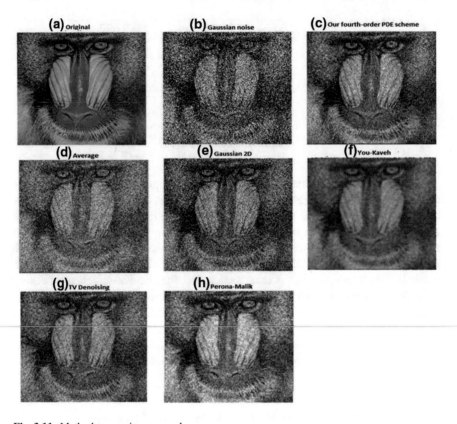

Fig. 3.11 Method comparison example

fourth-order diffusion-based model is derived from this variational scheme, using the corresponding Euler-Lagrange equation, which is:

$$\nabla^2\big(\psi_2^u\big(\big|\nabla^2 u\big|\big)\nabla u\big) - div\big(\psi_1^u(\|\nabla u\|)\nabla u\big) + \lambda(u - u_0) = 0 \qquad (3.112)$$

where $\psi_1^u, \psi_2^u : [0, \infty) \to (0, \infty)$, $\psi_1^u(s) = \frac{1}{s}\frac{\partial\varphi_1(s)}{\partial s}$, $\psi_2^u(s) = \frac{1}{s}\frac{\partial\varphi_2(s)}{\partial s}$.

Next, by applying the steepest descent method, we get the following PDE model with homogeneous Neumann boundary conditions:

$$\begin{cases} \frac{\partial u}{\partial t} = div\big(\psi_1^u(\|\nabla u\|)\nabla u\big) - \nabla^2\big(\psi_2^u\big(\big|\nabla^2 u\big|\big)\nabla u\big) - \lambda(u - u_0) \\ u(0, x, y) = u_0 \\ u(t, x, y) = 0, \text{ on } \partial\Omega \\ \frac{\partial u}{\partial \vec{n}} = 0 \end{cases} \qquad (3.113)$$

where the monotonic decreasing diffusivity functions are constructed as:

$$\begin{cases} \psi_1^u(s) = \begin{cases} \dfrac{\xi}{\left(\dfrac{s}{k(u)}\right)^2 + k(u)\left|\log_{10}\left(\dfrac{s}{k(u)}\right)\right|}, & \text{if } s > 0 \\ 1, & \text{if } s = 0 \end{cases} \\ \psi_2^u(s) = \alpha\sqrt{\dfrac{\eta(u)}{\beta s + \gamma}}, & \text{if } s \geq 0 \end{cases} \tag{3.114}$$

where $\xi, \beta, \gamma \in (1, 8]$, $\lambda, \alpha \in (0, 1)$ and the conductance parameters are modeled as:

$$\begin{cases} k(u) = \delta\mu(\|\nabla u\|) + rt(u) \\ \eta(u) = \left|median\left(|\nabla^2 u|\right) - 4\right| \end{cases} \tag{3.115}$$

where $\delta \in (2, 3)$ and $r \in (0, 1)$ [62].

So, the obtained nonlinear PDE contains a second-order anisotropic diffusion component and a fourth-order diffusion-based term. The second-order component assures that image filtering is performed along the gradient direction, the image details being preserved and the blurring avoided. The fourth-order component provides a more natural restored image and overcomes the staircase effect. Thus, the hybrid model benefits from the advantages of both types of diffusion.

The restored image is obtained by solving this compound fourth-order PDE model. So, the validity of this model is carefully investigated in [62], where the existence and uniqueness of a weak solution for (3.113), under some certain assumptions, is demonstrated. First, we have proven easily that the diffusivity functions are properly constructed for the diffusion process [62].

Then, a rigorous mathematical treatment of the PDE model's well-posedness has been performed [62]. Thus, this model has a weak solution if the flux functions $s\psi_1^u(s)$ and $s\psi_2^u(s)$ are monotonically increasing, which means their derivatives have to be positive. And because $\frac{\partial}{\partial s}\left(s\psi_i^u(s)\right) = \psi_i^u(s) + s\frac{\partial\psi_i^u(s)}{\partial s}$, $i \in \{1, 2\}$, it results that the following conditions have to be satisfied:

$$\begin{cases} \psi_1^u(s) + s\dfrac{\partial\psi_1^u(s)}{\partial s} \geq 0 \\ \psi_2^u(s) + s\dfrac{\partial\psi_2^u(s)}{\partial s} \geq 0 \end{cases}, \quad \forall s \geq 0 \tag{3.116}$$

We prove in [62], by performing the computations related to (3.116), that these conditions are verified for $s > k(u) \geq \ln(10)$. The unique weak solution that exists under these assumptions is determined by constructing a finite difference-based numerical approximation scheme. The same grid and the space and time coordinates given by (2.3) are used in this case, too. The second part of the PDE, which is related to the Laplacian operator, is approximated as follows:

$$f_1^n(i, j) = \Delta\left(\psi_2^u\left(|\nabla^2 u^n(i, j)|\right)\nabla^2 u^n(i, j)\right) \tag{3.117}$$

where $\nabla^2 u^n(i, j)$ is computed as in (3.104).

Since $div\big(\psi_1^u(\|\nabla u\|)\nabla u\big) = \psi_1^u(\|\nabla u\|)\Delta u + \nabla\big(\psi_1^u(\|\nabla u\|)\big) \cdot \nabla u$, the first part of the sum is approximated as $f_2^n(i, j) = \psi_1^u(|\nabla u^n(i, j)|)\nabla^2 u^n(i, j)$ and the second part can be written as:

$$
\big(\psi_1^u(\nabla u)\big) \cdot \nabla u = \psi_1^{u'}(\|\nabla u\|)\left(\frac{\frac{\partial u}{\partial x}\frac{\partial^2 u}{\partial x^2} + \frac{\partial u}{\partial y}\frac{\partial^2 u}{\partial x\partial y}}{\sqrt{\left(\frac{\partial u}{\partial x}\right)^2 + \left(\frac{\partial u}{\partial y}\right)^2}}, \frac{\frac{\partial u}{\partial y}\frac{\partial^2 u}{\partial y^2} + \frac{\partial u}{\partial x}\frac{\partial^2 u}{\partial x\partial y}}{\sqrt{\left(\frac{\partial u}{\partial x}\right)^2 + \left(\frac{\partial u}{\partial y}\right)^2}}\right)\left(\frac{\partial u}{\partial x}, \frac{\partial u}{\partial y}\right)
$$

$$
= \psi_1^{u'}(\|\nabla u\|)\frac{\left(\frac{\partial u}{\partial x}\right)^2\frac{\partial^2 u}{\partial x^2} + \frac{\partial u}{\partial x}\frac{\partial u}{\partial y}\frac{\partial^2 u}{\partial x\partial y} + \left(\frac{\partial u}{\partial y}\right)^2\frac{\partial^2 u}{\partial y^2} + \frac{\partial u}{\partial x}\frac{\partial u}{\partial y}\frac{\partial^2 u}{\partial x\partial y}}{\sqrt{\left(\frac{\partial u}{\partial x}\right)^2 + \left(\frac{\partial u}{\partial y}\right)^2}}
$$

$$(3.118)$$

If we consider that the second order derivatives do not vary so much, it is then approximated as:

$$
\nabla\big(\psi_1^u(\|\nabla u\|)\big) \cdot \nabla u \approx \psi_1^{u'}(\|\nabla u\|)\frac{\frac{\partial^2 u}{\partial x\partial y}\left(\frac{\partial u}{\partial x} + \frac{\partial u}{\partial y}\right)^2}{\sqrt{\left(\frac{\partial u}{\partial x}\right)^2 + \left(\frac{\partial u}{\partial y}\right)^2}}
$$

$$
\approx \psi_1^{u'}\left(\sqrt{\left(\frac{\partial u}{\partial x}\right)^2 + \left(\frac{\partial u}{\partial y}\right)^2}\right)\frac{\partial^2 u}{\partial x\partial y}\left(\frac{\partial u}{\partial x} + \frac{\partial u}{\partial y}\right) \quad (3.119)
$$

Then, the approximation result in (3.119) is discretized, by using the central differences, as $f_3^n(i, j)$. So, we get the explicit numerical approximation scheme:

$$
u_{i,j}^{n+1} = u_{i,j}^n(1 - \lambda) + \lambda u_{i,j}^0 + f_2^n(i, j) + f_3^n(i, j) - f_1^n(i, j) \quad (3.120)
$$

where we note $u^n(i, j) = u_{i,j}^n$ and also use the following boundary conditions:

$$
u^n(-1, j) = u^n(0, j), u^n(I + 1, j) = u^n(I, j), u^n(i, -1) = u^n(i, 0), u^n(i, J + 1) = u^n(i, J)
$$

The iterative numerical approximation algorithm given by (3.120) is consistent to the compound variational model (3.111) and converges fast to its solution. It has been successfully applied in our restoration experiments performed on hundreds of images corrupted with various levels of additive noise.

The algorithm has a low execution time, of approximately 1 s, and a high effectiveness. It removes a high amount of Gaussian noise, while preserving well the image edges and other details. Given its hybrid character, involving diffusions of different orders, it successfully overcomes both the blurring and the staircase effect. The optimal denoising results have been obtained for these parameter values:

Table 3.11 Average PSNR values

Method	Hybrid model	P-M 1	P-M 2	TV Denoising	You-Kaveh	Gaussian filter
PSNR (dB)	28.34	26.23	24.91	21.88	27.19	21.37

$$\xi = 1.5, \alpha = 0.7, \beta = 8, \gamma = 4, \lambda = 0.04, \delta = 2.4, r = 0.03, N = 25.$$

The proposed combined fourth-order PDE-based technique outperforms both the classic and PDE-based smoothing approaches. It produces better restoration results than conventional filters, by overcoming the blurring and preserving the features. It also outperforms the second-order PDE-based approaches, such as both versions of the Perona-Malik model and the ROF-TV scheme, achieving a better restoration and avoiding the staircase effect. The combined anisotropic diffusion technique performs also better than fourth-order isotropic diffusion-based methods, like You-Kaveh model, providing better detail preservation and avoiding the multiplicative noise. See some method comparison results in Table 3.11, which registers average PSNR values, and Fig. 3.12, which describes a *Barbara* image corrupted by Gaussian noise and restored by various techniques.

A weighted variant of this variational restoration model can also be considered. Some weight parameters corresponding to the two diffusions may be introduced in (3.111), to control the behavior of the hybrid framework, as follows:

$$J(u) = \int_{\Omega} \left(w_1 \varphi_1(\|\nabla u\|) + w_2 \varphi_2(\|\nabla^2 u\|) + \frac{\lambda}{2}(u - u_0)^2 \right) d\Omega \qquad (3.121)$$

Various enhancement effects could be achieved by varying the weighting coefficients $w_1, w_2 \in [0, 1]$. Increasing the w_1 value while decreasing w_2 would provide more weight to the second-order diffusion, leading to better deblurring, but also to more staircasing. Doing the opposite, would give more weight to the fourth-order diffusion, which means less blocky effect, but possible more detail over-filtering. The restoration algorithm can be applied repeatedly on the observed image, with various properly selected pairs $[w_1, w_2]$, to achieve an improved output.

Another compound fourth-order PDE-based denoising technique developed by us is combining a nonlinear fourth-order diffusion to a two-dimension Gaussian filter and a despeckling algorithm [63]. It contains also some boundary conditions, including a Neumann boundary condition, having the following form:

$$\begin{cases} \frac{\partial u}{\partial t} + \Delta\big(\delta_u(\|\nabla G_t * u\|)\nabla^2 u\big) + \beta(u - u_0) = 0 \\ u(0, x, y) = u_0(x, y), \forall (x, y) \in \Omega \\ u(t, x, y) = 0, \forall t \geq 0, (x, y) \in \partial\Omega \\ \Delta u(t, x, y) = 0, \forall (x, y) \in \partial\Omega \\ \frac{\partial u}{\partial n} = 0, \text{ on } \partial\Omega \end{cases} \qquad (3.122)$$

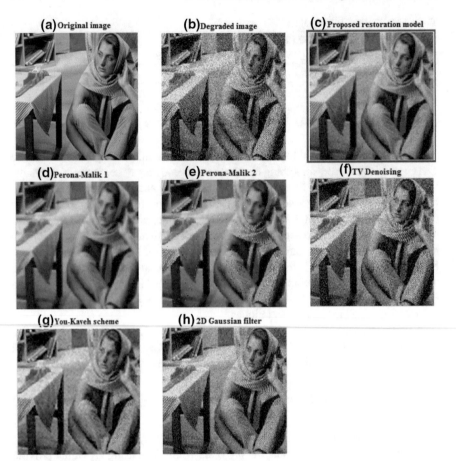

Fig. 3.12 Method comparison example

where $\beta \in [0, 1)$, $\Omega \subseteq R^2$, u_0 is the input image and the 2D Gaussian filter kernel $G_t(x, y) = \frac{1}{4\pi t} e^{-\frac{x^2+y^2}{4t}}$. The diffusivity function is modeled as:

$$\delta_u : [0, \infty) \to (0, \infty), \delta_u(s) = \frac{\nu}{\left(\frac{s}{\xi(u)}\right)^k + \xi(u)|\ln(s)|^{k-1}}, \qquad (3.123)$$

where $\nu \in (0, 4]$, $k \geq 2$ and $\xi(u) = \left| \frac{\mu(\|\Delta u\|) + median(\|\Delta u\|)}{\gamma} \right|$, $\gamma \in (1, 3]$.

The positive and monotonically decreasing function δ_u is properly chosen for the diffusion process. The nonlinear fourth-order PDE model is well-posed, admitting a unique weak solution that is computed using a finite difference-based approximation algorithm combined to a despeckling scheme [63].

The same quantization as in the previous case is used here, too. The PDE in (3.122) is discretized by applying the central differences on its components. So, the component $\frac{\partial u}{\partial t} + \beta(u - u_0)$ is approximated as:

$$\frac{u_{i,j}^{n+\Delta t} - u_{i,j}^n}{\Delta t} + \beta\left(u_{i,j}^n - u_{i,j}^0\right) = \frac{u_{i,j}^{n+\Delta t} + (\beta\Delta t - 1)u_{i,j}^n - \beta\Delta t u_{i,j}^0}{\Delta t} \tag{3.124}$$

The term $\Delta\left(\delta_u(\|\nabla G_t * u\|)\nabla^2 u\right)$ is discretized as:

$$\Delta\left(\varphi_{i,j}^n\right) = \frac{\varphi_{i+h,j}^n + \varphi_{i-h,j}^n + \varphi_{i,j+h}^n + \varphi_{i,j-h}^n - 4\varphi_{i,j}^n}{h^2} \tag{3.125}$$

where

$$\varphi_{i,j}^n = \delta_u\left(\nabla G_t * u_{i,j}^n\right)\frac{u_{i+h,j}^n + u_{i-h,j}^n + u_{i,j+h}^n + u_{i,j-h}^n - 4u_{i,j}^n}{h^2} \tag{3.126}$$

If we take $h = 1$ and $\Delta t = 1$, the following explicit numerical approximation scheme is achieved:

$$u_{i,j}^{n+1} = (1 - \beta)u_{i,j}^n - \varphi_{i+1,j}^n - \varphi_{i-1,j}^n - \varphi_{i,j+1}^n - \varphi_{i,j-1}^n + 4\varphi_{i,j}^n + \beta u_{i,j}^0 \tag{3.127}$$

that is applied on the evolving image, for n from 0 to N, i from 0 to I and j from 0 to J, but in combination to a despeckling algorithm.

Since fourth-order diffusions may generate speckle noise, we have constructed a scheme that removes this multiplicative noise, and integrate it into the iterative approximation algorithm [63]. At each step, if $u_{i,j}^{n+1}$ represents a speckle pixel (absolute difference between its value and the mean of its neighborhood is much greater than standard deviation of the neighborhood), it is replaced to its neighborhood' median. This despeckling process can be expressed as:

$$\left|u_{i,j}^{n+1} - \mu\left(u_{i,j}^{n+1}\right)\right| > \alpha\sigma\left(u_{i,j}^{n+1}\right) \Rightarrow u_{i,j}^{n+1} := median\left[N\left(u_{i,j}^{n+1}\right)\right] \tag{3.128}$$

where $\alpha > 1$ and $N\left(u_{i,j}^{n+1}\right) = u^{n+1}[i - r : i + r, j - r : j + r], r \in \{1, 2, 3\}$.

The resulted combined restoration approach removes both additive and multiplicative noise, while preserving details and overcoming the staircase effect. We have determined by trial and error method the values of the coefficients assuring an optimal image smoothing: $\beta = 0.1, \nu = k = 3, \gamma = r = 2, N = 17, \alpha = 2.7$.

Our compound fourth-order PDE filter outperforms both conventional and PDE-based approaches, achieving higher PSNR values, as one can see in Table 3.12. A method comparison example is also described in Fig. 3.13.

Table 3.12 PSNR values obtained by several techniques

Filtering technique	PSNR (dB)
Compound 4th-order PDE model	27.05
Average filter	25.31
Gaussian filter	25.24
Median filter	25.68
Perona-Malik scheme	25.98
TV-ROF Denoising	26.14
You-Kaveh algorithm	26.84

Fig. 3.13 Denoising results produced by some filters on *Elaine* image

3.4 Variational Denoising Frameworks Based on Nonlinear Control Schemes

We have also developed a class of variational restoration techniques based on non-linear control problems. In [64] we construct such effective denoising methods by solving nonconvex optimal control problems with the state and controller connected on a manifold described by a nonlinear elliptic equation.

The existence of a control that is able to restore a deteriorated image is demonstrated in [64] and the optimality conditions are determined by a passing to the limit method involving the Legendre-Fenchel relations. The type of image restoration approach proposed in that work is based on the next approximating control problem, which involves the mimimization of a differentiable cost functional:

$$\min_{(u,y,z)\in U_1} \int_\Omega \frac{|u|^q}{q} + \frac{\lambda}{2}(y - y^{obs})^2 + \frac{1}{\varepsilon}\Big(j(\nabla y) + j^*(\nabla z) - \nabla y \cdot \nabla z\Big) d\Omega \quad (3.129)$$

where $q \geq 2, \lambda \geq 0, j : R^N \rightarrow R$ is a convex differentiable function and j^* represents its conjugate, y^{obs} is the observed image, y is the evolving image obtained by the action of the control u and

$$U_1 = \left\{ \begin{array}{l} (u, y, z); u \in L^q(\Omega), y \in W^{1,1}(\Omega) \cap L^q(\Omega), z \in H^1(\Omega), j(\nabla y).j^*(\nabla z), \\ \nabla y \cdot \nabla z \in L^1(\Omega) \\ \int_\Omega u\, dx = 0, -\Delta z = u \text{ on } \Omega, \nabla z \cdot v = 0 \text{ on } \partial\Omega, \int_\Omega z\, dx = 0 \end{array} \right\}$$

$$(3.130)$$

Several hyphoteses are assumed on the function j: the weakly coercivity, symmetry at infinity, positiveness, differentiable conjugate. Also, the functions j and j^* satisfy the Legendre-Fenchel relations [64]. These properties make them appropriate for the image restoration process, being essential for the edge preserving [64].

Then, we demonstrate in [64] that the control problem (3.129) has at least a solution (u^*, y^*, z^*). Some optimality conditions are also provided for this problem [64]. Thus, if $\big(u_\varepsilon^*, y_\varepsilon^*, z_\varepsilon^*\big) \in U_1$ represents an optimal state for the problem (3.129), such that $a^{-1}(\nabla z^*) \in \big(L^2(\Omega)\big)^N$, these conditions of optimality are:

$$\big\|u_\varepsilon^*\big\|^{q-2}u_\varepsilon^* = \frac{1}{\varepsilon}y_\varepsilon^* - p_\varepsilon^* + C_\varepsilon^* \quad (3.131)$$

where C_ε^* represents a constant and the three variables are solutions to the following equations:

Table 3.13 Average PSNR values of several restoration methods

Model	Proposed method	Gaussian filter	Average filter	Wiener 2D	Perona-Malik	ROF-TV	You-Kaveh
PSNR (dB)	26.87	24.39	25.09	25.48	25.92	26.15	26.37

$$\begin{cases} \Delta p_\varepsilon = \frac{1}{\varepsilon} \nabla \cdot a^{-1}(\nabla z_\varepsilon^*) \text{ in } \Omega \\ \Delta p_\varepsilon \cdot v = \frac{1}{\varepsilon} \nabla \cdot a^{-1}(\nabla z_\varepsilon^*) \cdot v \text{ on } \partial\Omega \\ -\nabla \cdot a(\nabla y_\varepsilon^*) + \lambda\varepsilon(y_\varepsilon^* - y^{obs}) = u_\varepsilon^* \text{ in } \Omega \\ a(\nabla y_\varepsilon^*) \cdot v = 0 \text{ on } \partial\Omega \\ -\Delta z_\varepsilon^* = u_\varepsilon^* \text{ in } \Omega \\ -\Delta z_\varepsilon^* \cdot v = 0 \text{ on } \partial\Omega \end{cases} \tag{3.132}$$

where we have considered the following a function [64]:

$$a(r) = \log(|r|_N + 1) sign\, r + \frac{r}{|r|_N + 1} \tag{3.133}$$

A rigorous mathematical proof has been provided for these optimality conditions [64]. Then, a numerical approximation algorithm that solves the proposed variational model is constructed using these conditions.

So, the optimal state of the model, $(u_\varepsilon^*, y_\varepsilon^*, z_\varepsilon^*)$, is computed numerically from the equations in (3.131) and (3.132), by using the iterative steepest descent algorithm [64]. The discretization of these equations is described in detail in that paper [64]. Each of the three components, u^n, y^u and $z^u (n = 1, \ldots, N)$ is approximated iteratively using the steepest descent formula. The numerical approximation scheme converges fast to the optimal restoration solution, after a quite low number of iterations.

This obtained algorithm has been successfully tested on numerous degraded images, satisfactory results being produced. Our control-based variational method removes the Gaussian noise and preserves essential image features, such as edges and corners. It operates quite fast, given its fast-converging character, overcome the image blurring and reduces the staircase effect.

Method comparison have been also performed, the PSNR similarity metric being used to assess the performance of the approaches. The proposed technique outperforms the 2D conventional filters (Average, Gaussian, Wiener and others) and also some state-of-the-art second and fourth-order PDE-based restoration models, such as Perona-Malik scheme, total variation-based denoising and the You-Kaveh algorithm.

The average PSNR values achieved by the mentioned techniques are displayed in Table 3.13. The restoration results produced by these filtering methods on the noisy *Baboon* image ($\mu = 0.9$ and *var* = 0.07) are depicted in Fig. 3.14.

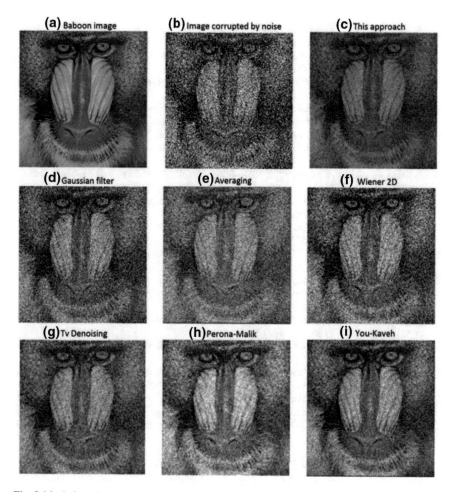

Fig. 3.14 *Baboon* image restored by this control-based approach and other methods

References

1. P. Perona, J. Malik, Scale-space and edge detection using anisotropic diffusion, in *Proceedings of the IEEE Computer Society Workshop on Computer Vision*, 16–22 November (1987)
2. R. Gonzalez, R. Woods, *Digital Image Processing*, 2nd edn. (Prentice Hall, 2001)
3. J. Weickert, *Anisotropic Diffusion in Image Processing*, European Consortium for Mathematics in Industry (B. G. Teubner, Stuttgart, 1998)
4. G. Aubert, P. Kornprobst, *Mathematical Problems in Image Processing: Partial Differential Equations and the Calculus of Variations*, vol. 147 (Springer Science & Business Media, 2006)
5. L. Rudin, S. Osher, E. Fatemi, Nonlinear total variation based noise removal algorithms. Physica D **60**(1), 259–268 (1992)
6. P. Charbonnier, L. Blanc-Feraud, G. Aubert, M. Barlaud, Two deterministic half-quadratic regularization algorithms for computed imaging, in *Proceedings of the IEEE International Conference on Image Processing*, vol. 2 (IEEE Comp. Society Press, Austin, TX, 1994), pp. 168–172

7. J. Weickert et al., Efficient and reliable schemes for nonlinear diffusion filtering. IEEE Trans. Image Process. **7**, 398–410 (1998)
8. M. Black, G. Shapiro, D. Marimont, D. Heeger, Robust anisotropic diffusion. IEEE Trans. Image Process. **7**(3), 421–432 (1998)
9. X. Li, T. Chen, Nonlinear diffusion with multiple edginess thresholds. Pattern Recogn. **27**(8), 1029–1037 (1994)
10. F. Voci, S. Eiho, N. Sugimoto, H. Sekiguchi, Estimating the gradient threshold in the Perona-Malik equation. IEEE Signal Process. Mag. **21**(3), 39–46 (2004)
11. D. Gleich, *Finite Calculus: A Tutorial for Solving Nasty Sums* (Stanford University, 2005)
12. P. Johnson, *Finite Difference for PDEs* (School of Mathematics, University of Manchester, Semester I, 2008)
13. F. Catte, P.L. Lions, J.M. Morel, T. Coll, Image selective smoothing and edge detection by nonlinear diffusion. SIAM J. Numer. Anal. **29**, 182–193 (1992)
14. J. Kacur, K. Mikula, Solution of nonlinear diffusion appearing in image smoothing and edge detection. Appl. Numer. Math. **17**, 47–59 (1995)
15. F. Torkamani-Azar, K.E. Tait, Image recovery using the anisotropic diffusion equation. IEEE Trans. Image Process. **5**, 1573–1578 (1996)
16. L. Alvarez, P.L. Lions, J.M. Morel, Image selective smoothing and edge detection by nonlinear diffusion II. SIAM J. Numer. Anal. **29**, 845–866 (1992)
17. C.A. Segall, S.T. Acton, Morphological anisotropic diffusion, in *Proceedings of the 1997 IEEE International Conference on Image Processing*, 26–29 October (Santa Barbara, CA, 1997), pp. 348–351
18. G. Gilboa, Y.Y. Zeevi, N.A. Sochen, Complex diffusion processes for image filtering. Lect. Notes Comput. Sci. **2106**, 299–307 (2001)
19. H. Yu, C.S. Chua, GVF-based anisotropic diffusion models. IEEE Trans. Image Process. **15**(6), 1517–1524 (2006)
20. O. Ghita, P.F. Whelan, A new GVF-based image enhancement formulation for use in the presence of mixed noise. Pattern Recogn. **43**, 2646–2658 (2010)
21. Y. Wang, Y. Jia, External force for active contours: gradient vector convolution, in *Pacific Rim International Conference on Artificial Intelligence* (2008), pp. 466–472
22. Y. Wang, W. Ren, H. Wang, Anisotropic second and fourth order diffusion models based on convolutional virtual electric field for image denoising. Comput. Math. Appl. **66**(10), 1729–1742 (2013)
23. M. Hazawinkel, *Variational Calculus*, Encyclopedia of Mathematics (Springer, 2001)
24. G. Arfken, The method of steepest descents, §7.4 in *Mathematical Methods for Physicists*, 3rd edn. (Academic Press, Orlando, FL, 1985), pp. 428–436
25. T.F. Chan, S. Esedoğlu, Aspects of total variation regularized L1 function approximation. SIAM J. Appl. Math. **65**(5), 1817–1837 (2005)
26. T. Le, R. Chartrand, T. Asaki, A variational approach to constructing images corrupted by Poisson noise. JMIV **27**(3), 257–263 (2007)
27. A. Buades, B. Coll, J.M. Morel, The staircasing effect in neighborhood filters and its solution. IEEE Trans. Image Process. **15**(6), 1499–1505 (2006)
28. P. Getreuer, Rudin–Osher–Fatemi total variation denoising using split Bregman. Image Process. On Line (2012)
29. C.A. Micchelli, L. Shen, Y. Xu, X. Zeng, Proximity algorithms for the L1/TV image denoising model. Adv. Comput. Math. **38**(2), 401–426 (2013)
30. Q. Chen, P. Montesinos, Q. Sun, P. Heng, D. Xia, Adaptive total variation denoising based on difference curvature. Image Vis. Comput. **28**(3), 298–306 (2010)
31. Y. Hu, M. Jacob, Higher degree total variation (HDTV) regularization for image recovery. IEEE Trans. Image Process. **21**(5), 2559–2571 (2012)
32. J. Yan, W.-S. Lu, Image denoising by generalized total variation regularization and least squares fidelity. Multidimension. Syst. Signal Process. **26**(1), 243–266 (2015)
33. Y.L. You, M. Kaveh, Fourth-order partial differential equations for noise removal. IEEE Trans. Image Process. **9**, 1723–1730 (2000)

34. H. Wang, Y. Wang, W. Ren, Image denoising using anisotropic second and fourth order diffusions based on gradient vector convolution. Comput. Sci. Inf. Syst. **9**, 1493–1512 (2012)
35. M. Lysaker, A. Lundervold, X.C. Tai, Noise removal using fourth-order partial differential equation with applications to medical magnetic resonance images in space and time. IEEE Trans. Image Process. **12**, 1579–1590 (2003)
36. T.F. Chan, A. Marquina, P. Mulet, High-order total variation based image restoration. SIAM J. Sci. Comput. **22**, 503–516 (2000)
37. S. Kim, H. Lim, Fourth-order partial differential equations for effective image denoising. Electron. J. Differ. Equ. Conf. **17**, 107–121 (2009)
38. T. Liu, Z. Xiang, Image restoration combining the second order and fourth-order PDEs. Math. Probl. Eng. **2013**(743891), 7 pages (2013)
39. F. Li, C.M. Shen, J.S. Fan, C.L. Shen, Image restoration combing a total variational filter and a fourth-order filter. J. Vis. Commun. Image Represent. **18**(4), 322–330 (2007)
40. O.M. Lysaker, X.-C. Tai, Iterative image restoration combining total variation minimization and a second-order functional. Int. J. Comput. Vision **66**, 5–18 (2006)
41. J. Wang, A noise removal model combining TV and a fourth PDE filter. Journal of Image and Graphics **13**(8) (2008)
42. V.S. Prasath, P. Kalavathi, Mixed noise removal using hybrid fourth order mean curvature motion, in *Advances in Signal Processing and Intelligent Recognition Systems* (Springer International Publishing, 2016), pp. 625–632
43. J. Rajan, K. Kannan, M.R. Kaimal, An improved hybrid model for molecular image denoising. J. Math. Imaging Vis. **31**, 73–79 (2008)
44. A.B. Hamza, P.L. Escamilla, J.M. Aroza, R. Roldan, Removing noise and preserving details with relaxed median filters. J. Math. Imaging Vis. 161–177 (1999)
45. T. Barbu, A. Favini, Rigorous mathematical investigation of a nonlinear anisotropic diffusion-based image restoration model. Electron. J. Differ. Equ. **2014**(129), 1–9 (2014)
46. T. Barbu, A. Ciobanu, C. Niță, Nonlinear second-order partial differential equation-based image smoothing technique. Mem. Sci. Sect. Rom. Acad. **XXXIX**, 7–14 (2016)
47. T. Barbu, A nonlinear parabolic partial differential equation model for image enhancement. Int. J. Comput. Inf. Eng. **3**(8) (2016)
48. T. Barbu, C. Morosanu, Image restoration using a nonlinear second-order parabolic PDE-based scheme. Analele Stiintifice ale Universitatii Ovidius Constanta, Seria Matematică **XXV**(1), 33–48 (2017)
49. T. Barbu, A novel variational PDE technique for image denoising, in *Proceedings of the 20th International Conference on Neural Information Processing, ICONIP 2013. Lecture Notes in Computer Science*, vol. 8228, part III, Daegu, Korea, 3–7 November, ed by M. Lee et al. (Springer, Berlin, 2013), pp. 501–508
50. T. Barbu, PDE-based Image Restoration using Variational Denoising and Inpainting Models, in *Proceedings of the 18th International Conference on System Theory, Control and Computing, ICSTCC 2014*, Sinaia, Romania, 17–19 October (2014), pp. 694–697
51. T. Barbu, Nonlinear PDE model for image restoration using second-order hyperbolic equations. Numer. Funct. Anal. Optim., Taylor & Francis **36**(11), 1375–1387 (2015)
52. T. Barbu, A nonlinear second-order hyperbolic diffusion scheme for image restoration. U.P.B. Sci. Bull., Ser. C **78**(2) (2016)
53. V. Barbu, *Nonlinear Semigroups and Differential Equations in Banach Spaces* (Noordhoff International Publishing, 1976)
54. J.M. Greenberg, R.C. MacCamy, V.J. Mizel, On the existence, uniqueness and stability of solutions of the equation $\sigma'(u_x)u_{xx} + \lambda u_{xtx} = \varrho_0 u_{tt}$. J. Math. Mech. **17**, 707–727 (1968)
55. E. Silva, K.A. Panetta, S.S. Agaian, Quantify similarity with measurement of enhancement by entropy, in *Proceedings: Mobile Multimedia/Image Processing for Security Applications, SPIE Security Symposium 2007*, vol. 6579, April (2007), pp. 3–14
56. T. Barbu, A nonlinear fourth-order PDE-based image denoising technique, in *Proceedings of the 23rd International Conference on Systems, Signals and Image Processing, IWSSIP 2016*, Bratislava, Slovakia, 23–25 May (IEEE, 2016), pp. 177–180

57. T. Barbu, Nonlinear fourth-order diffusion-based model for image denoising, in *Soft Computing Applications. Advances in Intelligent Systems and Computing*, vol. 633, ed. by V. Balas, L. Jain, M. Balas (Springer, Cham, 2017), pp. 423–429

58. A. Pazy, *Semigroups of Linear Operators and Applications to Partial Differential Equations* (Springer, Berlin, 1983)

59. T. Barbu, I. Munteanu, A nonlinear fourth-order diffusion-based model for image denoising and restoration. Proc. Rom. Acad., Ser. A: Math. Phys. Tech. Sci. Inf. Sci. **18**(2), 108–115 (2017)

60. A.P. Witkin, Scale-space filtering, in *Proceedings of the 8th International Joint Conference on Artificial Intelligence, IJCAI '83*, vol. 2, Karlsruhe, 8–12 August (1983), pp. 1019–1022

61. T. Barbu, Nonlinear fourth-order diffusion-based image restoration scheme. ROMAI J., ROMAI Soc. **1**, 1–8 (2016)

62. T. Barbu, PDE-based restoration model using nonlinear second and fourth order diffusions. Proc. Rom. Acad., Ser. A: Math. Phys. Tech. Sci. Inf. Sci. **16**(2), 138–146 (2015)

63. T. Barbu, A hybrid nonlinear fourth-order PDE-based image restoration approach, in *Proceedings of 20th International Conference on System Theory, Control and Computing ICSTCC 2016*, Sinaia, Romania, 13–15 October (IEEE, 2016), pp. 761–765

64. T. Barbu, G. Marinoschi, Image denoising by a nonlinear control technique. Int. J. Control, Taylor & Francis **90**(5), 1005–1017 (2017)

Chapter 4
Variational and PDE Models for Image Interpolation

The nonlinear PDE-based image reconstruction domain is approached in this chapter. The state of the art inpainting approaches based on differential models are described in the first section. The interpolation techniques based on variational models are described first and the PDE-based inpainting methods that do not follow variational principles are discussed next. Then, our main contributions in this image processing field are presented in the following sections. The structural inpainting techniques developed by us in variational or PDE form are based on second and fourth order nonlinear diffusion models. Our variational reconstruction algorithms are described in the second section, while the nonlinear PDE-based image interpolation solutions developed by us are detailed in the last section of this chapter.

4.1 State-of-the-Art PDE-Based Image Inpainting Techniques

Digital image *reconstruction*, also known as image *interpolation, inpainting* or *completion*, represents the process of recovering the missing or highly damaged regions of the image as plausibly as possible, using the information obtained from the known surrounding areas. The term inpainting originates from the ancient art restoration, where it is also called *retouching*. So, it represents a very old term, since the medieval artwork started to be restored as early as the Renaissance period [1].

While in the museum world the inpainting task is performed by an art restorer, the computer-based digital inpainting refers to the application of some interpolation procedures to replace the missing or corrupted image zones. The most important applications of the digital image completion include the following: damaged digital painting (artwork) reconstruction, photo and movie renovation, unwanted object removal, image super-resolution and zooming, and image (de)/compression [1].

© Springer International Publishing AG, part of Springer Nature 2019
T. Barbu, *Novel Diffusion-Based Models for Image Restoration and Interpolation*,
Signals and Communication Technology,
https://doi.org/10.1007/978-3-319-93006-0_4

The image reconstruction approaches are divided into the next three categories: textural inpainting, structural inpainting and combined techniques that perform both structure and texture-based interpolation. Some texture-based completion techniques are related to the texture synthesis [2, 3], while other approaches represent exemplar-based methods [4, 5]. Numerous texture inpainting techniques have been developed since an influential texture synthesis algorithm was introduced by Efros and Leung [2].

Structure-based inpainting employs information around the damaged zone to estimate isophote from coarse to fine, and diffuses information by diffusion mechanism. It uses energy-based and PDE-based models to reconstruct the missing or highly deteriorated image parts [6]. These models that follow the isophote directions in the image to perform the completion process are described in the next two subsections.

4.1.1   Variational Structure-Based Reconstruction Algorithms

The energy-based, or variational, structural inpainting techniques reconstruct the damaged image by solving a minimization problem involving an energy functional composed of a fidelity term and a regularizing term. Thus, a general variational inpainting model is formulated mathematically as follows:

$$\min_{u \in U} \left(J(u) = R(u) + \frac{1}{2} \int_{\Omega} \lambda_D (u - u_0)^2 d\Omega \right) \qquad (4.1)$$

where the image domain $\Omega \subseteq R^2$, D represents the inpainting domain whose *mask* is given by $\lambda_D = \lambda \cdot 1_{\Omega \backslash D}$, $\lambda > 0$ u_0 is the observed image and $R(u)$ represents the *regularizing term*, or the *prior image model*, and contains certain a priori information from the image u. The role of the fidelity term $\frac{1}{2} \int_{\Omega} \lambda_D (u - u_0)^2 d\Omega$ is to force the minimizer u to remain close enough to the initial image u_0 outside of the inpainting domain, while the minimization of the regularizing function is responsible for the filling in process [6].

If the regularizer term takes the form $R(u) = \int_{\Omega} \nabla u^2 dx dy$, then the Harmonic Inpainting model is obtained [1]. It represents a simple variational image interpolation approach that has some important drawbacks. Thus, Harmonic Inpainting does not satisfy the *connectivity principle*, since it cannot interpolate properly along the image gaps, and provides too smooth results. See the image a) in the inpainting example from Fig. 4.1.

The early variational inpainting solutions were based on the Mumford-Shah image model [1, 7]. The Mumford-Shah based image interpolation is based on the next minimization problem:

$$\min_u \left(\frac{1}{2} \int_{\Omega} \lambda_D (u - u_0)^2 dx dy + \frac{\gamma}{2} \int_{\Omega \backslash \Gamma} |\nabla u|^2 dx dy + \alpha H^1(\Gamma) \right) \qquad (4.2)$$

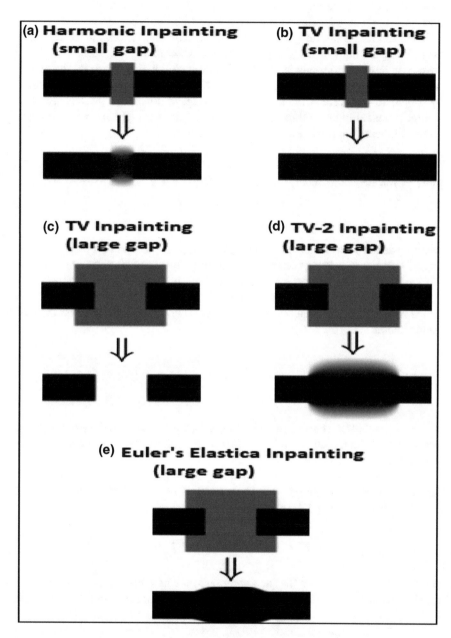

Fig. 4.1 Connectivity results produced by variational methods

where Γ represents the edge set of the image and $H^1(\Gamma)$ is the one-dimension Hausdorff measure that generalizes the length notion for regular curves.

The Mumford-Shah image model has some shortcomings that make it insufficient for inpainting [7]. So, numerous improved variational inpainting algorithms have been derived from this model, by approximating the edge set and the length. So, the Γ- convergence approximation of the Mumford-Shah functional, introduced by Ambrosio and Tortorelli in the context of image segmentation [8], has been successfully applied to inpainting [7]. The obtained image completion model is based on the following minimization:

$$\min_u \left(\frac{1}{2} \int_\Omega \lambda_D (u - u_0)^2 d\Omega + \frac{\gamma}{2} \int_{\Omega \backslash \Gamma} z_\in^2 |\nabla u|^2 d\Omega + \alpha \int_\Omega \left(\in |\nabla z_\in|^2 + \frac{(1 - z_\in)^2}{4 \in} \right) d\Omega \right)$$

(4.3)

where $z_\in$ denotes the signature function of the edge set. The inpainting scheme based on (4.3) has a low order of complexity in terms of approximation and computation. Also, it has a rapid numerical convergence and preserves the sharpness of the edges.

Another improved inpainting technique is based on the Mumford-Shah-Euler image model [1, 6, 7]. That variational interpolation scheme is based on the next minimization:

$$\min_u \left(\frac{1}{2} \int_\Omega \lambda_D (u - u_0)^2 dxdy + \frac{\gamma}{2} \int_{\Omega \backslash \Gamma} |\nabla u|^2 dxdy + \int_\Gamma (\alpha + \beta \kappa^2) ds \right) \quad (4.4)$$

where κ represents the curvature, ds the length element and the coefficients $\alpha, \beta > 0$ An elliptic PDE completion model is obtained by applying the corresponding Euler-Lagrange equation on (4.4).

An influential variational image inpainting technique is the Total Variation (TV) Inpainting scheme introduced by T. Chan and J. Shen in 2001 [1, 9]. It is based on the TV prior image model, having the following form:

$$\min_{u \in L^2(\Omega)} \left(J[u] = \int_\Omega \alpha \|\nabla u\| dxdy + \frac{\lambda_D}{2} \int_\Omega (u - u_0)^2 dxdy \right) \quad (4.5)$$

where $\alpha > 0$ D represents the inpainting region, where the image information is missing, u_0 constitutes the observed image and the Lagrange multiplier is $\lambda_D = \lambda \cdot (1 - 1_D)$ 1_D being the characteristic function of the inpainting domain.

Thus, the TV Inpainting model fills in the missing regions by minimizing the first-order total variation while keeping close to the observed image in the known regions. The parameter α controls the smoothing process. If a low value for it is chosen, then the smoothing is directed mainly to the inpainting zone and is minimal outside of the region D.

The following second-order nonlinear anisotropic diffusion-based model is then derived from (4.5), by applying its Euler-Lagrange equation and the steepest descent method:

$$
\begin{cases}
\frac{\partial u}{\partial t} = \alpha div\left(\frac{\nabla u}{|\nabla u|}\right) - \lambda_D(u - u_0) \\
u(0, x, y) = u_0
\end{cases}
\tag{4.6}
$$

The obtained PDE-based inpainting scheme works successfully for non-texture type images affected by missing or highly damaged zones. The TV Inpainting approach achieves connectivity, but not for the large gaps (see Fig. 4.1b, c). Also, it may generate the undesired image staircasing, since the total variation minimization produces piecewise constant structures.

Since the image interpolation may be viewed as filtering with a spatially-varying regularization strength λ_D, the TV regularization-based inpainting model is closely related to the ROF − TV Denoising scheme given by (3.17)–(3.18). Because of their similarity, the TV Denoising with Split Bregman algorithm [10] has been easily adapted to the inpainting domain [11].

As in the total variation based denoising case, many TV-based reconstruction approaches that improve the TV Inpainting technique have been proposed in the last years, most of them having higher orders. One of them is the TV^2 Inpainting model, which is obtained by considering the second-order total variation as a regularizer term: $R(u) = \int_\Omega |\nabla^2 u| dxdy$ [1, 12].

TV^2 Inpainting outperforms TV Inpainting, achieving better interpolation results. It produces more natural reconstructed images, being able to overcome the blocky effect. Also, TV^2 Inpainting technique represents a better connectivity solution, providing a satisfactory interpolation along large gaps, as one can see in Fig. 4.1d.

Some variational algorithms that combine the first- and the second-order TV regularizations achieve improved image interpolation results. Such a $TV + TV^2$ Inpainting framework is described in [13]. The image corrupted by missing regions is recovered by that combined total variation − based technique, as following:

$$
u^* = \underset{u}{\arg\min}\left(\frac{\lambda_D}{2}\int_{\Omega \setminus D} (u - u_0)^2 dxdy + \int_\Omega \alpha(x)|\nabla u| dxdy + \int_\Omega \beta(x)|\nabla^2 u| dxdy\right)
\tag{4.7}
$$

where u^* represents the reconstructed image, D is the inpainting zone, $\alpha(x)$ and $\beta(x)$ are two properly modeled spatially varying functions.

The compound variational approach given by (4.7) provides effective image completion results, outperforming both TV and TV^2 Inpainting algorithms. It interpolates successfully large inpainting regions, avoiding the creation of undesired blocky-like structures in the recovered images. It is numerically solved by employing the Split Bregman algorithm [11, 13].

A generalized version of the TV model, which is the Total Generalized Variation (TGV) that involves higher-order derivatives of u and is successfully used for image denoising, can be also applied for inpainting [14]. The TGV model has the form:

$$TGV_\alpha^2(u) = \sup\left\{\int_\Omega u \operatorname{div}^2 v \, dx dy : v \in C_c^2\left(\Omega, Sym^2\left(R^d\right)\right), \operatorname{div}^l v_\infty \le \alpha_l, l = 0, 1\right\} \quad (4.8)$$

where $Sym^2\left(R^d\right)$ represents the space of symmetric tensors of second order with arguments in R^d. Several image interpolation techniques using this TGV scheme have been proposed in [15, 16]. Other important total variation-based reconstruction frameworks are Blind Inpainting using l_0 and TV Regularization [17] and TV Inpainting with Primal-Dual Active Set (PDAS) method [18].

Another influential higher-order variational interpolation framework developed by T. Chan and J. Shen is Euler's Elastica Inpainting model [19]. It is based on an earlier variational inpainting algorithm using Euler's Elastica in the context of disocclusion, introduced in 1998 by Masnou and Morel [20], and is obtained from (4.1) by applying the next regularizer:

$$R(u) = \int_D w(u)\left(\alpha + \beta\left(\nabla \cdot \frac{\nabla u}{|\nabla u|}\right)^2\right)|\nabla u| dx dy \quad (4.9)$$

where the parameters $\alpha, \beta > 0$ control the behaviour of this interpolation model and $w(u)$ represents a weighting function depending on the evolving image's histogram.

Their approach provides an effective reconstruction of the damaged image, being able to inpaint large missing regions and working properly in noisy conditions, too. Obviously, Euler's Elastica Inpainting provides a much better connectivity than TV Inpainting model, being able to interpolate along much larger image gaps [19]. One can see an example of inpainting result produced by Euler's Elastica in Fig. 4.1e.

We have also elaborated many such variational structural inpainting approaches and disseminated them in some of our past papers. These research contributions are described in the following sections.

4.1.2 Nonlinear Second and High-Order PDE Inpainting Models

Also, numerous second and higher order PDE-based inpainting techniques have been developed in the last two decades. Usually, the second-order differential models for image interpolation follow variational principles, which means those PDE schemes can be derived from the variational inpainting approaches. Unlike them, the higher-order PDE inpainting approaches are not derived from some variational problems, being directly provided as evolutionary differential equations.

Since the inpainting and restoration constitute two strongly related image processing fields, the second-order PDE-based denoising algorithms can be easily adapted for the image reconstruction. The general form of a nonlinear second-order PDE inpainting model is obtained by making the Frechet derivative of the functional J at u equal to zero and applying the steepest descent method next. So, we get:

$$\frac{\partial u}{\partial t} + \nabla R(u) + \lambda_D(u - u_0) = 0 \; in \; \Omega \tag{4.10}$$

where $\lambda_D = \lambda \cdot (1 - 1_D)$ and $\nabla R(u)$ represents here the Frechet derivative of the regularizing function. Depending on the selection of this regularizing term R, various second-order PDE-based image inpainting approaches have been developed in the last decades.

Numerous important state of the art interpolation frameworks do not come from variational principles, being based on higher-order partial differential equations. They include some influential third- and fourth-order PDE inpainting models.

Thus, the third order PDE-based inpainting technique introduced by M. Bertalmio, G. Sapiro, C. Ballester and V. Caselles in 2000 represents a pioneering work in this domain [21]. The *digital inpainting* terminology first appeared in their paper that disseminated a reconstruction model based on observations about old artwork restoration [21].

The proposed PDE-based interpolation algorithm, replicating the basic approaches used by art restorers, propagates the necessary information (image Laplacian) in the direction of the isophotes (boundaries) [21]. It is characterized by the following equation:

$$\frac{\partial u}{\partial t} = \nabla^\perp u \cdot \nabla \Delta u \tag{4.11}$$

where $\nabla^\perp u$ is the perpendicular gradient of the image u.

This nonlinear third-order PDE provided by (4.11) is numerically solved inside the inpainting region D. It represents a transport equation for the image smoothness that is modeled by the image's Laplacian along the level lines of that image. The additional anisotropic diffusion introduced to avoid the level-line crossing leads to the next PDE:

$$\frac{\partial u}{\partial t} = \nabla^\perp u \cdot \nabla \Delta u + \nabla \cdot (g(|\nabla u|)\nabla u) \tag{4.12}$$

where $\nu > 0$. The goal is to evolve this PDE to a steady-state solution, where $\nabla^\perp u \cdot \nabla \Delta u = 0$ which ensures that information is constant in the direction of the isophotes.

In [22] the authors show a relation between this interpolation method and the Navier–Stokes equations for fluid dynamics. The current research in the Navier-Stokes equation-based inpainting field is mostly inspired by their completion approach [23]. Also this third-order PDE inpainting model aims to prove that both the gradient direction and gray-scale values have to be propagated inside inpainting domain, and shows why high-order PDEs are needed for a proper reconstruction.

The inpainting algorithm developed by Bertalmio et al. reconstructs successfully the static and video images affected by missing (damaged) zones and provides a very good text and object removal. A successful inpainting example from [21] is displayed in Fig. 4.2, which describes an old photograph reconstructed by this method.

The Curvature-driven Diffusion (CDD) Inpainting is another well-known third-order PDE-based structural reconstruction technique elaborated by Chan and Shen [1, 24]. They developed CDD Inpainting model as a solution to fix the drawbacks of their TV Inpainting scheme. The proposed reconstruction model, which uses the curvature information of the isophotes (level lines), is characterized by the following form:

$$
\begin{cases}
\frac{\partial u}{\partial t} = \nabla \cdot \left(\frac{g(\kappa)}{|\nabla u|} \nabla u \right) - \lambda_D (u - u_0) \\
u(0, x, y) = u_0
\end{cases}
\tag{4.13}
$$

where the curvature of the isophote is $\kappa = \nabla \cdot \left[\frac{\nabla u}{|[\nabla u]|} \right]$ and g is a positive and continuous function chosen as $g(s) = s^p$, $p \geq 1$ [24].

The CDD Eq. (4.13) diffuses the smoothness perpendicularly to the level lines. It preserves the direction of these level lines and is able to connect the level lines across large distances. That makes CDD Inpainting a better reconstruction solution than TV Inpainting and other second-order PDE-based interpolation methods.

However, the CDD Inpainting model has its own disadvantages. Thus, it has also a noise sensitive character, not performing properly in noisy image conditions.

A finite difference-based numerical algorithm that solves the PDE in (4.13) is provided in [24]. The flux of this curvature-driven diffusion model is computed from (2.1) to (2.2) and (4.13) as:

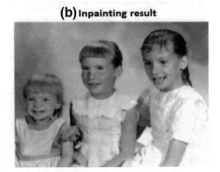

Fig. 4.2 Old photo reconstructed by the inpainting model of Bertalmio et al.

$$j = -\frac{g(\kappa)}{|\nabla u|} \nabla u \qquad (4.14)$$

Then, the CDD equation, which can be written as $\frac{\partial u}{\partial t} = -\nabla \cdot j$ is discretized using the next explicit numerical approximation scheme:

$$u^{n+1} = u^n - \Delta t [\nabla \cdot j]^{(n)} \qquad (4.15)$$

where the sampling $t = n\Delta t$ and $[\nabla \cdot j]^{(n)}$ represents the discretization of the divergence $\nabla \cdot j$ which is obtained by using the half-point central differences for the divergence operator [24]. The iterative scheme (4.15) converges stable to the inpainting result.

CDD Inpainting technique performs effectively various interpolation tasks, such as disocclusions, reconstruction of old photos affected by scratches, text removal from images and image object removal from scene. A CDD Inpainting based text removal example is described in Fig. 4.3 [24].

Another important category of high-order PDE-based inpainting approaches is that of fourth-order PDE interpolation models. One of them is the Cahn-Hilliard Inpainting scheme [1, 25].

Cahn-Hilliard equation represents a nonlinear fourth-order diffusion equation originating in the material science, but the respective inpainting model is based on a modified version of this PDE, a fidelity term being added to it. Thus, if $u_0 \in L^2(\Omega)$ represents the observed image, affected by missing regions, its inpainted version is obtained from the following evolution equation with Neumann boundary conditions:

$$\begin{cases} u_t = -\Delta\left(\varepsilon\Delta u - \frac{1}{\varepsilon}W'(u)\right) - \lambda(x, y)(u - u_0), \, in \, \Omega \\ \frac{\partial u}{\partial v} = \frac{\partial \Delta u}{\partial v} = 0, \, on \, \partial\Omega \end{cases} \qquad (4.16)$$

(a) Original image

(b) Inpainted image

(c) Inpainting mask

Fig. 4.3 Text removal result produced by CDD Inpainting

where the nonlinear double-well potential function $W(u) = u^2(u-1)^2$, and $\lambda(x, y) =$
$$\begin{cases} \lambda_0 \geq 1, (x, y) \in \Omega \backslash D \\ 0, (x, y) \in D \end{cases} \text{where } D \text{ is the inpainting domain.}$$

The nonlinear fourth-order PDE model (4.16) is well-posed, the global existence of a unique weak solution of it being demonstrated in [25]. The solution is computed by using a consistent, fast converging and stable numerical approximation scheme. So, the Cahn-Hilliard Inpainting model is solved numerically by applying the *convexity splitting* fast solver [1, 25].

The convexity splitting approach divides the energy functional of the PDE into two parts: a convex energy and a concave one. Then, the component of the Euler-Lagrange equation derived from the convex part is treated implicitly, while the component derived from the concave part is treated explicitly in the numerical approximation scheme [25].

Cahn-Hilliard Inpainting performs an effective image interpolation that outperform the nonlinear second-order PDE-based technique. It provides a smooth continuation of level lines into the inpainting domain D, like the curvature-based inpainting algorithms, while converging faster than CDD Inpainting technique [25].

The reconstruction of a cross using Cahn-Hilliard Inpainting is displayed in Fig. 4.4 [25]. See the observed image with the square inpainting region in (a) and the inpainting result after 1000 iterations in (b).

Another fourth-order PDE-based inpainting technique is $TV - H^{-1}$ Inpainting scheme [1, 26]. It is described as following:

$$u_t = \Delta p + \lambda(x, y)(u_0 - u), p \in \partial TV(u) \qquad (4.17)$$

where the observed image $u_0 \in L^2(\Omega)$ $\lambda(x, y)$ is defined as in the previous case and $\partial TV(u)$ represents the subdifferential of

$$TV(u) = \begin{cases} |Du|(\Omega), & if |u| \leq 1 \, a.e. \, in \, \Omega \\ +\infty, otherwise \end{cases} \qquad (4.18)$$

with $|Du|(\Omega)$ representing the total variation of u.

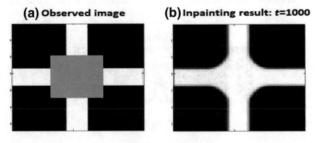

(a) Observed image   **(b) Inpainting result: t=1000**

Fig. 4.4 Text removal result produced by CDD Inpainting

It is demonstrated that the stationary equation $\Delta p + \lambda(u_0 - u) = 0$ admits a solution $u \in BV(\Omega)$ [1]. That means the inpainting Eq. (4.17) has a steady state.

The $TV - H^{-1}$ Inpainting model is solved numerically by using the following time-stepping numerical approximation scheme:

$$\frac{U_{k+1} - U_k}{\Delta t} + C_1 \Delta\Delta U_{k+1} + C_2 U_{k+1} =$$
$$C_1 \Delta\Delta U_k - \Delta\left(\nabla \cdot \left(\frac{\nabla U_k}{|\nabla U_k|_\epsilon}\right)\right) + C_2 U_k + \lambda(u_0 - U_k) \tag{4.19}$$

where $C_1 > \frac{1}{\epsilon}$ and $C_2 > \lambda_0$.

The $TV - H^{-1}$ Inpainting provides an effective image reconstruction, outperforming the second-order TV Inpainting model, but requires a high number of iterations, given the complexity of its approximation scheme (4.19). For example, the observed image u_0 displayed in Fig. 4.5 (a) is inpainted successfully by this fourth-order PDE model after $t = 1000$ iterations, with the parameters $\lambda_0 = 10^2$, $\epsilon = 0.01$ the interpolation result being displayed in the image (b) [1].

The inpainting with *low curvature image simplifiers* (LCIS) represents another fourth-order PDE-based image interpolation solution [1, 27]. Proposed in 1999 and inspired by the Perona-Malik anisotropic diffusion scheme, LCIS Inpainting model is characterized by the following nonlinear fourth-order PDE:

$$u_t = -\nabla \cdot (g(|\nabla u|)\nabla \Delta u) + \lambda(u_0 - u) \tag{4.20}$$

where $u_0 \in L^2(\Omega)$ $\lambda(x, y)$ is given as in the previous cases and the function $g(s) = \frac{1}{1+s^2}$ The Eq. (4.20) leads to

$$u_t = -\nabla \cdot (\arctan(\Delta u) + \lambda(u_0 - u) \tag{4.20}$$

which represents a gradient flow in L^2 for the energy functional

(a) Observed image

(b) Inpainting after 1000 iterations

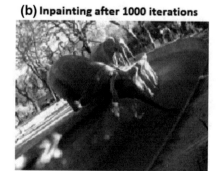

Fig. 4.5 $TV - H^{-1}$ Inpainting: $\lambda_0 = 10^2$, $\epsilon = 0.01$

(a) Observed image   **(b) Inpainting result**

Fig. 4.6 LCIS Inpainting: $\lambda_0 = 10^3, t = 1000$

$$E(u) = \int_\Omega G(\Delta u)dxdy + \frac{1}{2}\int_\Omega \lambda(u_0 - u)^2 \qquad (4.21)$$

where $G'(s) = \arctan(s)$. According to the convexity splitting approach, this energy functional is written as $E(u) = E_1(u) - E_2(u)$ where

$$\begin{cases} E_1(u) = \int_\Omega \frac{C_1}{2}(\Delta u)^2 dxdy + \frac{1}{2}\int_\Omega \frac{C_2}{2}|u|^2 dxdy \\[2mm] E_2(u) = \int_\Omega -G(\Delta u) + \frac{C_1}{2}(\Delta u)^2 dxdy + \frac{1}{2}\int_\Omega -\lambda(u_0 - u)^2 + \frac{C_2}{2}|u|^2 dxdy \end{cases} \qquad (4.22)$$

and the constants C_1 and C_2 are chosen such that E_1 and E_2 are all strictly convex.

Then, one obtains the following time-stepping scheme that solves numerically the PDE inpainting model:

$$\frac{U_{k+1}-U_k}{\Delta t} + C_1\Delta^2 U_{k+1} + C_2 U_{k+1} = \qquad (4.23)$$
$$-\Delta(\arctan \Delta U_k) + C_1\Delta^2 U_k + \lambda(u_0 - U_k) + C_2 U_k$$

LCIS Inpainting produces successful image interpolation results, but because of the computational expensive numerical approximation algorithm (4.23), it does not execute fast. See an LCIS inpainting example described in Fig. 4.6, where a text is removed from the observed image after a high number of iterations of this algorithm [1].

We have elaborated many nonlinear PDE-based inpainting models in the recent years. The most important of our contributions in this area are discussed in the Sect. 4.3.

4.2 Variational Interpolation Solutions for Non-textured Images

We have conducted a high amount of research in the structural inpainting domain, developing numerous energy and PDE based image interpolation techniques in the last decade. These reconstruction methods have been disseminated in many articles published in recognized international journals or volumes of international conferences.

In this section we describe our main contributions in the energy-based (variational) inpainting domain. Some of these variational interpolation techniques lead to nonlinear second-order diffusion-based models. They are described in the next subsection. We have also developed a class of compound variational inpainting models combining nonlinear second and fourth order diffusions, which is presented in the second subsection.

Consistent finite-difference method-based numerical approximation schemes and robust mathematical investigations are provided for the described variational approaches in the following subsections. The successful reconstruction experiments performed using these methods are also discussed.

4.2.1 Variational Inpainting Models Using Nonlinear Second-Order Diffusions

An effective class of variational frameworks that provide a successful inpainting, even in noisy conditions, and overcome the undesirable effects is proposed by us in [28].

The considered variational approach minimizes an energy cost functional involving an image inpainting mask. So, the recovered image is determined as $u_{opt} = \arg\min_u F(u)$ where

$$F(u) = \int_\Omega \left(\alpha \varphi_u(\nabla u) + \frac{\beta}{2}(1 - 1_\Gamma)(u - u_0)^2 \right) d\Omega \qquad (4.24)$$

where $\alpha, \beta \in (0, 1)$ the image domain $\Omega \subseteq R^2$, the inpainting domain $\Gamma \subset \Omega$ the inpainting mask is based on its characteristic function, $1_\Gamma(x, y) = \begin{cases} 1, & \forall(x, y) \in \Gamma \\ 0, & \forall(x, y) \notin \Gamma \end{cases}$ the observed image is a partial 2D function defined only outside of the missing zone, $u_0 : \Omega \backslash \Gamma \to R$ the evolving image $u : \Omega \to R$ satisfies $u|_{\Omega \backslash \Gamma} = u_0$ and the regularizer function is constructed as:

$$\varphi_u(s) = \int_0^s \tau \xi \left(\frac{\eta(u)}{\gamma \log_{10}(s + \eta(u))^2 + \delta} \right)^{\frac{1}{3}} d\tau \qquad (4.25)$$

whose conductance parameter is based on some statistics of the image u:

$$\eta(u(x, y, t)) = \zeta median(\|\nabla u\|) + vt \tag{4.26}$$

where $\xi, \gamma, \zeta, v \in (0, 1]$ and $\delta \in [1, 5)$ [28].

If we note $E(x, y, u, u_x, u_y) = \alpha\varphi_u(\nabla u) + \frac{\beta}{2}(1 - 1_\Gamma)(u - u_0)^2$, then the Euler-Lagrange equation corresponding to (4.24) is determined as:

$$\frac{\partial E}{\partial u} - \frac{\partial}{\partial x}\frac{\partial E}{\partial u_x} - \frac{\partial}{\partial y}\frac{\partial E}{\partial u_y} = 0 \tag{4.27}$$

leading to $\beta(1 - 1_\Gamma)(u - u_0) - \frac{\partial}{\partial x}\left(\alpha\varphi_u'(\nabla u)\frac{2u_x}{\nabla u}\right) - \frac{\partial}{\partial y}\left(\alpha\varphi_u'(\nabla u)\frac{2u_y}{\nabla u}\right) = 0$ that is equivalent to

$$\beta(1 - 1_\Gamma)(u - u_0) - 2\alpha div\left(\frac{\varphi_u'(\nabla u)}{\nabla u}\nabla u\right) = 0 \tag{4.28}$$

If we note $\psi_u(s) = \frac{\varphi_u'(s)}{s}$ the following positive and monotonically decreasing diffusivity function is obtained:

$$\psi_u : [0, \infty) \to (0, \infty), \quad \psi_u(s) = \xi\left(\frac{\eta(u)}{\gamma \log_{10}(s + \eta(u))^2 + \delta}\right)^{\frac{1}{3}} \tag{4.29}$$

So, Eq. (4.28) becomes $\beta(1 - 1_\Gamma)(u - u_0) - 2\alpha div(\psi_u(\|\nabla u\|)\nabla u) = 0$ and the steepest gradient descent method is then applied to it [29]. Thus, one obtains a nonlinear anisotropic diffusion model having the form:

$$\begin{cases} \frac{\partial u}{\partial t} = 2\alpha div(\psi_u(\|\nabla u\|)\nabla u) - \beta(1 - 1_\Gamma)(u - u_0) \\ u(0, x, y) = u_0 \\ u(t, x, y) = 0, \; on \; \partial\Omega\backslash\Gamma \end{cases} \tag{4.30}$$

A rigorous mathematical treatment of the validity of this PDE-based model is performed in [28], the existence of a unique weak solution being demonstrated. The anisotropic diffusion-based model (4.30) is well-posed, since its flux function, $s\psi_u(s)$ is monotonically increasing, having a positive derivative.

Since $(s\psi_u(s))' = \frac{\partial}{\partial s}(s\psi_u(s)) = \psi_u(s) + s\frac{\psi_u(s)}{\partial s}$ we get:

$$(s\psi_u(s))' = \xi\left(\frac{\eta(u)}{\gamma \log_{10}(s + \eta(u))^2 + \delta}\right)^{\frac{1}{3}} - \frac{2s\gamma\eta(u)^{\frac{1}{3}}\xi}{3\ln(10)(s + \eta(u))(2\gamma \log_{10}(s + \eta(u)) + \delta)^{\frac{4}{3}}}$$

$$= \frac{\xi\eta(u)^{\frac{1}{3}}}{(\gamma \log_{10}(s + \eta(u))^2 + \delta)^{\frac{1}{3}}}\left(1 - \frac{2s\gamma}{3\ln(10)(s + \eta(u))(2\gamma \log_{10}(s + \eta(u)) + \delta)}\right) > 0 \tag{4.31}$$

because $3 \ln(10)(s + \eta(u))(2\gamma \log_{10}(s + \eta(u)) + \delta) > 2s\gamma$ if $s + \eta(u) > 10$ a condition that is generally true, the conductance parameter taking quite high values [28]. The unique weak solution of the nonlinear PDE model (4.30) is converging to the solution of the variational scheme: $\lim_{t \to \infty} u(t, x, y) = u_{opt}$ [28]. That solution is computed using a numerical approximation algorithm that solves the PDE variational model. A stable and fast converging numerical discretization scheme is constructed in [28], by applying the finite difference method [30, 31].

A finite difference based discretization similar to that described in the Sect. 3.2.2, which corresponds to a variational restoration model, can be applied in this case. The same quantization of the space and time coordinates given by (2.13) is used here, too. By applying the discretization-related procedures provided by the Eqs. (3.56)–(3.63), one obtains the following explicit numerical approximation scheme:

$$u_{i,j}^{n+1} = u_{i,j}^{n}(1 - \beta(1 - 1_\Gamma)) + 2\alpha\psi_u\left(\frac{\sqrt{\left(u_{i+1,j}^{n} - u_{i-1,j}^{n}\right)^2 + \left(u_{i,j+1}^{n} - u_{i,j-1}^{n}\right)^2}}{2}\right)$$

$$\cdot \left(u_{i+1,j}^{n} + u_{i-1,j}^{n} + u_{i,j+1}^{n} + u_{i,j-1}^{n} - 4u_{i,j}^{n}\right)$$

$$+ 2\alpha\psi_u'\left(\frac{\sqrt{\left(u_{i+1,j}^{n} - u_{i-1,j}^{n}\right)^2 + \left(u_{i,j+1}^{n} - u_{i,j-1}^{n}\right)^2}}{2}\right)$$

$$\cdot \frac{\left(u_{i+1,j+1}^{n} - u_{i+1,j-1}^{n} - u_{i-1,j+1}^{n} + u_{i-1,j-1}^{n}\right)\left(u_{i+1,j}^{n} - u_{i-1,j}^{n} + u_{i,j+1}^{n} - u_{i,j-1}^{n}\right)}{8}$$

$$+ \beta(1 - 1_\Gamma)u_{i,j}^{0} \qquad\qquad (4.32)$$

The iterative numerical approximation algorithm (4.32) is applied on the evolving $[I \times J]$ image for $n = 0, \ldots, N, I = 1, \ldots, I$ and $J = 1, \ldots, J$. It is consistent to the nonlinear diffusion-based model (4.30) and converges fast to its solution, since the number of iterations required for an optimal image interpolation is quite low.

This variational inpainting approach has been tested on hundreds of images affected by missing parts. It fills in successfully these regions, preserving successfully image details, such as edges and corners, and works properly in noisy conditions, too, avoiding the unintended effects.

Given the fast convergence of its iterative approximation algorithm, this technique has a low execution time, of approximately 1 s. However, this running time and the number of iterations, N, depends on size of the inpainting domain, and the amount of noise.

The optimal inpainting results are achieved for some properly chosen parameters of this model. The values of these coefficients, detected by trial and error method, are $\alpha = 0.4, \beta = 0.4, \xi = 0.5, \gamma = 0.7, \delta = 4, \zeta = 1.4, v = 0.05, \Delta t = 1, h = 1$

Method comparison have been also performed by us. The performance of the proposed technique has been assessed using well-known measures, such as Peak Signal-to-Noise Ratio (PSNR) and Norm of the Error Image, and compared to other inpainting techniques. This interpolation technique outperforms many state of the art inpainting approaches, getting better average values of the performance measures. So, it provides better completion results and executes faster than TV Inpainting and Harmonic Inpainting models, and also performs better than reconstruction algorithms derived from second-order diffusion-based restoration schemes, such as the Perona-Malik model. Also, it outperforms inpainting methods based on Gaussian processes, such as the Gaussian Process Regression (GPR) model [32].

One can see the average PSNR values corresponding to various techniques in Table 4.1. Our technique achieves higher PSNR values than most inpainting approaches. It is outperformed by TV Inpainting using PDAS and provides slightly better values than Kriging Inpainting model [33].

Some text removal experiments, performed in both normal and noisy conditions, are described in Fig. 4.7. One can see the interpolation output produced by our variational scheme, TV Inpainting, Harmonic Inpainting and P-M based inpainting model after $t = 25$ iterations, first when applied on *Lenna* image affected only by missing part (b), and then when it is applied on *Lenna* image corrupted by both missing part and additive Gaussian noise characterized by $\mu = 0.21$ and *variance* = 0.02 (g). Obviously, the proposed technique achieves much better results in this time frame.

Another variational interpolation technique, from the same class provided by Eq. (4.24), has been developed by us and disseminated in [34]. It is based on other versions of regularizer, diffusivity function and conductance parameter, and provides effective inpainting results, while executing fast and avoiding the undesirable effects.

4.2.2 Variational Inpainting Combining Second-to Fourth-Order Diffusions

We have also constructed a class of hybrid variational inpainting frameworks based on a combination of nonlinear second and fourth order diffusions. The compound

Table 4.1 Average PSNR values achieved by some inpainting methods

Inpainting technique	Average PSNR value (dB)
This variational model	31.25
TV Inpainting	26.38
Harmonic Inpainting	23.54
P-M based Inpainting	26.38
Kriging Interpolation	30.38
TV with PDAS	33.96

Fig. 4.7 Text removal results obtained by several inpainting methods at the moment $t = 25$

variational model proposed in [35] is meant to perform a successful filling of the missing part while reducing the additive Gaussian noise and overcoming all the unintended effects.

The reconstructed image is dketermined by solving the following minimization problem, involving an energy cost functional that is based on two regularizer functions, which corresponds to the second-order and fourth-order nonlinear diffusions:

$$u_{inp} = \arg\min_u \int_\Omega \varphi_u^1(\nabla u) + \varphi_u^2(|\Delta u|) + \frac{\alpha(1 - 1_\Gamma)}{2}(u - u_0)^2 d\Omega \qquad (4.33)$$

where $\alpha \in (0, 1]$ u_{inp} represents the recovered image, u_0 is the observed image the image domain $\Omega \subseteq R^2$, the inpainting domain $\Gamma \subset \Omega$ and the two regularizers are positive functions, $\varphi_u^1, \varphi_u^2 : (0, \infty) \to (0, \infty)$ expressed as:

$$\varphi_u^1(s) = \int_0^s \tau\lambda\left(\frac{\eta_u}{\beta\tau^k + \xi}\right)^{\frac{1}{3}} d\tau$$

$$\varphi_u^2(s) = \int_0^s \tau\xi\sqrt{\frac{\eta_u}{\varepsilon \log_{10}(\eta_u + \tau)^3 + \gamma}} d\tau \qquad (4.34)$$

where $\lambda, \beta, \varepsilon \in (0, 1]$ $\xi, \gamma \in [1, 3]$ $k \in (0, 4)$ and the conductance parameter is modeled as:

$$\eta_u = |\nu\mu(\nabla u) - \zeta t| \qquad (4.35)$$

where the coefficients $\nu, \zeta \in (0, 1)$ [35].

The variational inpainting scheme (4.32) is solved by determining the nonlinear PDE model corresponding to it. So, one determines the Euler-Lagrange equation associated to that variational model. It has the form:

$$\frac{\partial J}{\partial u} - \frac{\partial}{\partial x}\frac{\partial J}{\partial u_x} - \frac{\partial}{\partial y}\frac{\partial J}{\partial u_y} = 0 \qquad (4.36)$$

where

$$J(x, y, u, u_x, u_y) = \varphi_u^1(\|\nabla u\|) + \varphi_u^2(|\Delta u|) + \frac{\alpha(1 - 1_\Gamma)}{2}(u - u_0)^2 \qquad (4.37)$$

From (4.36) one obtains:

$$\nabla^2\left(\frac{\varphi_u^{2\prime}(|\Delta u|)}{|\Delta u|}\Delta u\right) - div\left(\frac{\varphi_u^{1\prime}(\nabla u)}{\nabla u}\nabla u\right) + \alpha(u - u_0) = 0 \qquad (4.38)$$

Then, one introduces the following two positive diffusivity functions:

$$\delta_1^u, \delta_2^u : [0, \infty) \to (0, \infty), \delta_1^u(s) = \frac{\varphi_u^{1\prime}(s)}{s}, \delta_2^u(s) = \frac{\varphi_u^{2\prime}(s)}{s} \qquad (4.39)$$

and obtains:

$$\Delta\big(\delta_2^u(|\Delta u|)\Delta u\big) - div\big(\delta_1^u(\nabla u)\nabla u\big) + \alpha(1 - 1_\Gamma)(u - u_0) = 0 \qquad (4.40)$$

The steepest gradient descent method is then applied on (4.40) and the next non-linear fourth-order parabolic PDE results:

$$\frac{\partial u}{\partial t} - div\big(\delta_1^u(\|\nabla u\|)\nabla u\big) + \Delta\big(\delta_2^u(|\nabla^2 u|)\Delta u\big) + \alpha(1 - 1_\Gamma)(u - u_0) = 0 \quad (4.41)$$

The following combined fourth-order anisotropic diffusion-based model with boundary conditions is then obtained:

$$\begin{cases} \frac{\partial u}{\partial t} = div\big(\delta_1^u(\|\nabla u\|)\nabla u\big) - \Delta\big(\delta_2^u(|\nabla^2 u|)\Delta u\big) - \alpha(1 - 1_\Gamma)(u - u_0) \\ u(t, x, y) = 0, \; on \; \partial\Omega\backslash\Gamma \\ u(0, x, y) = u_0 \\ \frac{\partial u}{\partial n} = 0 \end{cases} \qquad (4.42)$$

Obviously, this hybrid PDE-based model is composed of two nonlinear diffusion terms. Besides providing an improved inpainting, these components remove the additive noise and alleviate different types of undesired effects. The second-order diffusion component overcome the blurring effect and assures a detail-preserving restoration. The nonlinear fourth-order diffusion-based term overcomes the unintended staircase effect.

The solution of the variational problem (4.32) is determined by solving the fourth-order diffusion model (4.42). Since this compound PDE model is well-posed because of its Dirichlet condition, a weak solution of this model exists and it is also unique [35].

That solution representing the inpainted image is computed numerically by using a consistent and fast-converging finite difference-based discretization algorithm. The proposed approximation scheme uses the same space grid size of h and the time step Δt and the same quantization of the time and space coordinates: $x = ih$, $y = jh$, $t = n\Delta t$, $\forall i \in \{0, 1, \ldots, I\}$, $j \in \{0, 1, \ldots, J\}$, $n \in \{0, 1, \ldots, N\}$

First, the second-order diffusion term, $div\big(\delta_1^u(\nabla u)\nabla u\big)$ is approximated as:

$$D_1 = \sigma \sum_{j \in N_p} \delta_1^u\big(\nabla_{p,q}(n)\big) \cdot \nabla_{p,q}(n) \qquad (4.43)$$

where $N_p = \{(i - 1, j), (i + 1, j), (i, j - 1), (i, j + 1)\}$ represents the four-neighborhood of the pixel $p = (i, j)$, $\sigma \in (0, 1)$ while the gradient magnitude in a particular direction at the time n is determined as $\nabla u_{p,q}(n) = u(q, n) - u(p, n)$

Then, the fourth-order diffusion-based component is approximated using the Laplacian discretization. So, we have:

$$\delta_{i,j}^u = \delta_2^u\big(|\nabla^2 u_{i,j}|\big)\Delta u_{i,j}^n \qquad (4.44)$$

where

$$\Delta u_{i,j}^n = \frac{u_{i+h,j}^n + u_{i-h,j}^n + u_{i,j+h}^n + u_{i,j-h}^n - 4u_{i,j}^n}{h^2} \qquad (4.45)$$

Then, one applies the discrete Laplacian on (4.44) and obtains the approximation:

$$D_2 = \Delta \delta_{i,j}^u = \frac{\delta_{i+1,j}^u + \delta_{i-1,j}^u + \delta_{i,j+1}^u + \delta_{i,j-1}^u - 4\delta_{i,j}^u}{h^2} \qquad (4.46)$$

The following implicit discretization is obtained for the nonlinear PDE-based model:

$$\frac{u_{i,j}^{n+\Delta t} - u_{i,j}^n}{\Delta t} = D_1 - D_2 - \alpha(1 - 1_\Gamma)\left(u_{i,j}^n - u_{i,j}^0\right) \qquad (4.47)$$

which leads to the next iterative explicit numerical approximation scheme, for $\Delta t = 1$

$$u_{i,j}^{n+1} = u_{i,j}^n(1 - \alpha(1 - 1_\Gamma)) + D_1 - D_2 + \alpha(1 - 1_\Gamma)u_{i,j} \qquad (4.48)$$

The explicit numerical approximation algorithm (4.48) is stable and consistent to the combined nonlinear diffusion-based interpolation framework. It converges fast to its solution, reaching the optimal image inpainting after a low number of iterations.

Many interpolation experiments have been conducted using the described compound technique, satisfactory reconstruction results being obtained. Some well-known image collections, such as the USC - SIPI database, have been used in our inpainting tests. The selection of the model's parameters is important for the interpolation output. These empirically identified coefficients lead to the optimal results:

$$\alpha = 0.5, \lambda = 0.4, \beta = 0.7, \varepsilon = 0.6, \nu = 2, k = 2, \zeta = 0.2, \gamma = 1.5, \xi = 1.7, h = 1, N = 25 \qquad (4.49)$$

The proposed compound technique executes fast, given its fast-converging character, and provide a proper structural reconstruction of the damaged images, completing successfully the missing regions, in both normal and noisy conditions. It preserves the edges and other features, and avoids the image blurring and staircasing effects. While our variational approach produce satisfactory results for non-textured images affected by missing parts, it cannot inpaint properly the image textures.

Structure-based interpolation method comparisons have been also performed by us. The performance of our interpolation technique has been assessed using PSNR and MSE similarity metrics.

The average PSNR values achieved by various inpainting techniques are registered in Table 4.2. The described algorithm achieves higher PSNR values than some well-known interpolation methods, such as the TV Inpainting model, and also it runs much faster than them.

The inpainting results produced by this variational approach and other PDE-based methods on a noisy *Baboon* image affected also by a missing zone (black scratch) are displayed in Fig. 4.8. One can see that our hybrid framework provides the best output.

Table 4.2 Average PSNR values of PDE inpainting models

Inpainting method	Average PSNR (dB)
The proposed variational model	31.95
TV Inpainting	29.96
Harmonic Inpainting	27.32
Mumford-Shah based Inpainting	30.27
Bertalmio et al. model	31.75

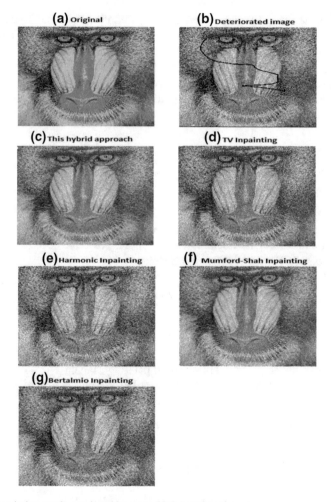

Fig. 4.8   Inpainting results produced by some PDE-based techniques

4.3 Nonlinear PDE-Based Structural Reconstruction Methods

In this section we describe our most important contributions in the structure-based inpainting domain using nonlinear PDE models. The nonlinear diffusion-based interpolation schemes developed by us have been disseminated in many publications. Some of our models follow variational principles and can be derived from energy-based approaches like those described in the previous section, while other PDE models have a non-variational character.

In the first subsection there are described the parabolic PDE-based inpainting algorithms proposed by us. A class of hyperbolic PDE interpolation models is described in the second subsection, while a more complex nonlinear anisotropic diffusion-based interpolation framework is presented in the last subsection.

Consistent numerical discretizations using finite differences are constructed for these nonlinear differential models. Rigorous mathematical treatments of some models are also provided.

4.3.1 Second-Order Parabolic Nonlinear PDE Models for Image Inpainting

A nonlinear second-order anisotropic diffusion-based model that provides an effective structural inpainting is disseminated in [36]. It is based on the following second-order parabolic PDE with boundary conditions:

$$\begin{cases} \frac{\partial u}{\partial t} = \nabla \cdot (\varphi^u(\|\nabla u\|)\nabla u) - \xi(1 - 1_\Gamma)(u - u_0) \\ u(0, x, y) = u_0 \\ u(t, x, y) = 0, \, on \, \partial\Omega\backslash\Gamma \end{cases} \tag{4.50}$$

where the coefficient $\xi \in (0, 1]$ and the inpainting domain $\Gamma \subset \Omega \subseteq R^2$. The diffusivity function of this nonlinear PDE-based model is constructed as follows:

$$\varphi^u : [0, \infty) \to (0, \infty), \, \varphi^u(s) = \alpha \sqrt[3]{\frac{\delta_u}{\beta(s + \gamma)^k + \lambda}} \tag{4.51}$$

where $\alpha, \beta \in (0, 1), \lambda, \gamma \in [1, 3]$ and the conductance parameter is modeled as:

$$\delta_u = |\nu median(\|\nabla u\|) + \zeta \mu(\|\nabla u\|)| \tag{4.52}$$

where $\nu, \zeta \in (0, 1)$

This nonlinear anisotropic diffusion model can also be obtained from a variational scheme. It reconstructs properly the damaged image by directing the diffusion mainly to the unknown regions. The diffusivity function is properly constructed for this diffusion process, being positive, monotonically increasing and converging to zero.

A rigorous mathematical investigation is performed on this second-order parabolic nonlinear PDE-based model in [36], its well-posedness being seriously analyzed. Thus, we demonstrate the existence and uniqueness of a weak solution for anisotropic diffusion model (4.50), if the condition

$$u \in C\big([0, T]; H_0^1(\Omega)\big) \cap L^2\big([0, T]; L^2(\Omega)\big) \tag{4.53}$$

is satisfied. That solution is then computed using an iterative numerical approximation algorithm.

Since the nonlinear partial differential equation from (4.50) can be re-written as

$$\frac{\partial u}{\partial t} = \varphi^u(|\nabla u|)\Delta u + \nabla\big(\varphi^u(|\nabla u|)\big) \cdot \nabla u - \xi(1 - 1_\Gamma)(u - u_0) \tag{4.54}$$

it is discretized by its components in (4.54).

The first one, $\frac{\partial u}{\partial t}$ is approximated using forward differences as $\frac{u_{i,j}^{n+\Delta t} - u_{i,j}^n}{\Delta t}$ The second component, $\varphi^u(|\nabla u|)\Delta u$ is discretized using central differences, as

$$D_1^n(i, j) = \varphi^u\big(|\nabla u_{i,j}^n|\big)\Delta u_{i,j}^n \tag{4.55}$$

where $\Delta u_{i,j}^n$ is given by (4.45). The third term, $\nabla(\varphi^u(|\nabla u|)) \cdot \nabla u$ gets the next central difference-based approximation

$$D_2^n(i, j) = \varphi^{u\prime}\big(|\nabla u_{i,j}^n|\big)$$
$$\frac{u_{i+h,j+h}^n - u_{i+h,j-h}^n - u_{i-h,j+h}^n + u_{i-h,j-h}^n}{4h^2} \frac{u_{i+h,j}^n - u_{i-h,j}^n + u_{i,j+h}^n - u_{i,j-h}^n}{2h} \tag{4.56}$$

where $i, j, n, h, \Delta t$ are given by (2.13). Obviously, the last component is approximated as $\xi(1 - 1_\Gamma)\big(u_{i,j}^n - u_{i,j}^0\big)$

We may consider $\Delta t = 1$ So, we get the following explicit numerical approximation scheme:

$$u_{i,j}^{n+1} = u_{i,j}^n(1 - \xi(1 - 1_\Gamma)) + D_1^n(i, j) + D_2^n(i, j) + \xi(1 - 1_\Gamma)u_{i,j}^0 \tag{4.57}$$

The iterative numerical approximation algorithm (4.57) is consistent to the parabolic model (4.50) and converges fast to its solution. It has been successfully used in our inpainting experiments [36].

Those interpolation experiments have recovered hundreds of images affected by missing regions, in both normal and noisy conditions [36]. The following set of empirically detected parameters provides the optimal image interpolation results:

$$\xi = 0.5, \beta = 0.4, \alpha = 0.6, \zeta = 1.3, \gamma = 1.4, k = 2, \nu = 3, h = 1, N = 25 \tag{4.58}$$

Method comparison are also described in [36]. This PDE-based inpainting approach outperforms some well-known PDE and variational completion approaches, such as TV Inpainting and Harmonic Inpainting models, performing better image reconstruction results and achieving better values of the performance measures [36].

See the average PSNR values achieved by several PDE models, which are registered in the following table. The parabolic PDE-based technique described here obtains the highest average value (Table 4.3).

In Fig. 4.9 the *Baboon* image corrupted by a missing part representing a black scratch is inpainted by the PDE-based reconstruction schemes from the above table. The images (c–f) illustrate the inpainting results produced by these methods after $N = 25$ iterations. One can see that our interpolation technique achieves the best output after that number of steps.

Another nonlinear parabolic PDE-based inpainting approach is described in [37]. It uses the next nonlinear anisotropic diffusion equation with boundary conditions:

$$\begin{cases} \frac{\partial u}{\partial t} - \alpha div(\psi_u(\|\nabla u\|)\nabla u) + \gamma(1 - 1_\Gamma)(u - u_0) + F \cdot \nabla u = 0 \\ u(0, x, y) = u_0 \\ u(t, x, y) = 0, \, on \, \partial\Omega \backslash \Gamma \end{cases} \tag{4.59}$$

where the coefficients $\alpha \in (0, 6] \, \gamma \in (0, 1]$ and the drift term is constructed as:

$$F(x, y) = \left(e^{-\lambda_1(x^k + y^k)}, e^{-\lambda_2(x^{k+1} + y^{k+1})}\right) \tag{4.60}$$

where $k, \lambda_1, \lambda_2 > 0$

The following positive diffusivity function is considered for this PDE model [37]:

$$\psi_u : [0, \infty) \to [0, \infty)$$

$$\psi_u(s) = \frac{\lambda}{\eta(u)\left|\ln\left(\frac{s}{\eta(u)}\right)\right|^k + \delta\left(\frac{s}{\eta(u)}\right)^2} \tag{4.61}$$

Table 4.3 The PSNR values obtained by several PDE models

Inpainting techniques	Average PSNR (dB)
The proposed PDE model	32.05
TV Inpainting	31.35
Harmonic Inpainting	27.52
P-M based Inpainting	29.86

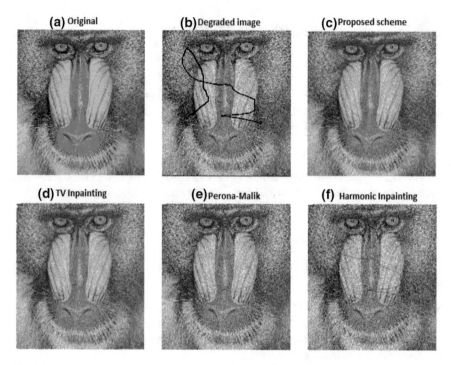

Fig. 4.9 Damaged *Baboon* image inpainted by several PDE models ($t = 25$)

where $\alpha, \delta \in (1, 3]$ and its conductance parameter is modeled using the evolving image's statistics, as:

$$\eta(u) = \left| \frac{\xi \mu(\nabla u) + \beta \cdot median(\nabla u)}{\nu} \right| \qquad (4.62)$$

where $\xi, \beta, \nu \in (0, 3]$ The function ψ_u is characterized by the main properties required by a successful diffusion process: positive, monotonically decreasing and converging to zero for ψ_u [37].

The nonlinear anisotropic diffusion-based process provided by (4.59)–(4.62) may be viewed as a restoration scheme that performs the interpolation by directing the filtering mostly to the missing regions and reducing it outside of them. This diffusion process is controlled by the selection of the model's parameters α and γ If the value of α is chosen small then the denoising is minimal outside of the inpainting domain. If γ receives a low value, which is close to 0, and the value of α is increased, then the proposed PDE-based model is transformed into a restoration scheme.

The reconstructed image is determined by solving this parabolic differential model. The model (4.59) is well-posed, accepting an unique and weak solution [37]. The solution is computed numerically using an iterative finite difference-based algorithm.

Table 4.4 The NE values obtained by several PDE-based methods

Inpainting approach	Norm of the error
The proposed parabolic PDE model	5.3×10^3
TV Inpainting	5.9×10^3
Harmonic	6.4×10^3
Mumford-Shah based Inpainting	5.4×10^3

So, a finite difference method based discretization similar to that in the previous case has been applied to this model [36, 37]. The next explicit numerical approximation scheme is obtained:

$$u_{i,j}^{n+1} = u_{i,j}^n (1 - \gamma (1 - 1_\Gamma)) + \alpha \left(D_1^u + D_2^u \right) + \gamma (1 - 1_\Gamma) u_{i,j}^0$$
$$- \left(e^{-\lambda_1 (i^k + j^k)}, e^{-\lambda_2 (i^{k+1} + j^{k+1})} \right) \left(\frac{u_{i+1,j}^n - u_{i-1,j}^n}{2}, \frac{u_{i,j+1}^n - u_{i,j-1}^n}{2} \right) \qquad (4.63)$$

where the discretizations D_1^u and D_2^u are determined as in (4.55) and (4.56) [36].

The iterative numerical algorithm (4.63), which is consistent to our parabolic model, produces an effective and fast inpainting of the deteriorated images when applied with the following parameters:

$$\gamma = 0.8, \alpha = 5.2, \alpha = 1.2, \lambda = 1.5, \lambda_1 = 2, \lambda_2 = 3, k = 2, \xi = 1.3, \beta = 1.4, \nu = 2, N = 28 \tag{4.64}$$

The proposed anisotropic diffusion approach provides better structural interpolation results than many existing PDE-based models, but performs considerably weaker for texture-based inpainting. One can see an inpainting example illustrating the performance of our technique in Fig. 4.10. The images in (c–f) illustrate the reconstruction results produced by our model and other approaches on the *Peppers* image affected by 2 missing zones, after $N = 28$ iterations.

The performance of the described technique has been assessed by various similarity measures. See the average Norm of the Error (NE) value achieved by our approach, which is better (lower) than the values obtained by the other methods, in Table 4.4.

4.3.2 Nonlinear Hyperbolic PDE-Based Structural Interpolation Schemes

A class of nonlinear second-order hyperbolic PDE-based techniques for structure-based inpainting has been proposed in [38]. It is based on the following second-order hyperbolic partial differential equations and several boundary conditions:

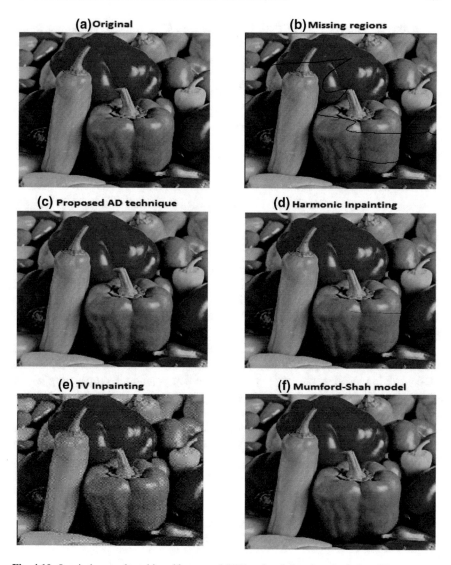

Fig. 4.10 Inpainting results achieved by several PDE and variational methods ($t = 28$)

$$
\begin{cases}
\gamma \frac{\partial^2 u}{\partial t^2} + \lambda^2 \frac{\partial u}{\partial t} - div(\xi_u(|\nabla u|) \cdot \nabla u) + \beta(1 - 1_M)(u - u_0) = 0 \\
u(0, x, y) = u_0(x, y) \\
\frac{\partial u}{\partial t}(0, x, y) = u_1(x, y) \\
u(t, x, y) = 0, \quad \forall (x, y) \in \partial\Omega \backslash M
\end{cases}
\tag{4.65}
$$

where the parameters $\gamma, \lambda, \beta \in (0, 1]$ and $M \subseteq \Omega \subseteq R^2$ represents the inpainting domain. Obviously, the evolving image is given as a function $u : \Omega \to R u|_{\Omega \setminus M} = u_0$. The edge-stopping function of this hyperbolic model is constructed as:

$$\xi_u : [0, \infty) \to [0, \infty)$$

$$\xi_u(s) = \varepsilon \left(\frac{K(u)}{\alpha \log_{10}(s + K(u))^k + v} \right)^{1/3} \tag{4.66}$$

where $\alpha, \varepsilon \in (0, 1)$ $v \in (1, 5]$ $k \in \{1, 2, 3, 4\}$ and the conductance parameter has the form:

$$K(u(x, y, t)) = |\eta median(\|\nabla u\|) - \delta t| \tag{4.67}$$

where $\eta \in (1, 3)$ and $\delta \in (0, 1)$

The diffusivity function provided by (4.66) is properly modeled, being always positive and monotonically decreasing, and it is converging to zero for $s \to \infty$. By adding the second-order time derivative of the evolving image, the proposed model removes the diffusion effect in the neighborhood of the edges and produces a strong filtering and detail-preserving effect, besides filling successfully the inpainting region.

The hyperbolic model given by (4.65) is also well-posed, accepting a unique and weak solution, under some certain assumptions. Also, this second-order differential model has the localization property [39], since that solution is propagating with a finite speed. The solution of this PDE, which represents the recovered digital image, is then determined numerically using a finite-difference method-based approach.

Thus, the same space grid size of h and the time step Δt and almost the same quantization of the time and space coordinates used in the previous cases has been used here too:

$$x = ih, y = jh, t = n\Delta t, i \in \{1, \ldots, I\}, j \in \{1, \ldots, J\}, n \in \{1, \ldots, N\} \tag{4.68}$$

The hyperbolic equation in (4.65) is divided into two components which are discretized using finite differences [30, 31]. The first discretization process is expressed as:

$$D\left(\gamma \frac{\partial^2 u}{\partial t^2} + \lambda^2 \frac{\partial u}{\partial t} + \beta(1 - 1_M)(u - u_0) \right) = \gamma \frac{u_{i,j}^{n+\Delta t} + u_{i,j}^{n-\Delta t} - 2u_{i,j}^n}{\Delta t^2}$$

$$+ \lambda^2 \frac{u_{i,j}^{n+\Delta t} - u_{i,j}^{n-\Delta t}}{2\Delta t} + \beta(1 - 1_M)\left(u_{i,j}^n - u_{i,j}^0 \right) \tag{4.69}$$

that leads to

$$D\left(\gamma \frac{\partial^2 u}{\partial t^2} + \lambda^2 \frac{\partial u}{\partial t} + \beta(1 - 1_M)(u - u_0)\right) = u_{i,j}^{n+\Delta t}\left(\frac{\gamma}{\Delta t^2} + \frac{\lambda^2}{2\Delta t}\right)$$
$$+ u_{i,j}^{n-\Delta t}\left(\frac{\gamma}{\Delta t^2} - \frac{\lambda^2}{2\Delta t}\right) + u_{i,j}^n\left(\beta(1 - 1_M) - \frac{2\gamma}{\Delta t^2}\right) + u_{i,j}^0 \beta(1_M - 1) \quad (4.70)$$

where $D(\ )$ symbolizes the discretization of the argument. If one considers $\Delta t = 1$ then (4.70) leads to

$$D\left(\gamma \frac{\partial^2 u}{\partial t^2} + \lambda^2 \frac{\partial u}{\partial t} + \beta(1 - 1_M)(u - u_0)\right) = u_{i,j}^{n+1}\left(\frac{2\gamma + \lambda^2}{2}\right)$$
$$+ u_{i,j}^{n-1}\left(\frac{2\gamma - \lambda^2}{2}\right) + u_{i,j}^n(\beta(1 - 1_M) - 2\gamma) + u_{i,j}^0 \beta(1_M - 1) \quad (4.71)$$

Next, the discretization of $div(\xi_u(\nabla u) \cdot \nabla u)$ component is performed as following:

$$D(div(\xi_u(\|\nabla u\|) \cdot \nabla u)) = \zeta \sum_{q \in V_p} \xi_u\left(\|\nabla^{p,q} u\|\right) \cdot \nabla^{p,q} u \quad (4.72)$$

where $\zeta \in (0, 0.5)$ $V_p = \{(i - h, j), (i + h, j), (i, j - h), (i, j + h)\}$ and we may consider the value $h = 1$. Obviously, the gradient magnitude in a particular direction at time n is computed as $\nabla^{p,q} u(n) = u(q, n) - u(p, n)$ [38].

So, the following implicit numerical approximation scheme is obtained for the hyperbolic PDE:

$$u_{i,j}^{n+1}\left(\frac{2\gamma + \lambda^2}{2}\right) + u_{i,j}^{n-1}\left(\frac{2\gamma - \lambda^2}{2}\right) + u_{i,j}^n(\beta(1 - 1_M) - 2\gamma)$$
$$- \zeta \sum_{q \in N_p} \xi_u\left(\nabla^{p,q} u\right) \cdot \nabla^{p,q} u + u_{i,j}^0 \beta(1_M - 1) = 0 \quad (4.73)$$

that leads to the next explicit numerical approximation algorithm:

$$u_{i,j}^{n+1} = u_{i,j}^n\left(\frac{4\gamma - 2\beta(1 - 1_M)}{2\gamma + \lambda^2}\right) + u_{i,j}^{n-1}\left(\frac{\lambda^2 - 2\gamma}{2\gamma + \lambda^2}\right) -$$
$$+ \zeta \sum_{q \in N_p} \xi_u\left(\nabla^{p,q} u\right) \cdot \nabla^{p,q} u + u_{i,j}^0 \beta(1 - 1_M) = 0 \quad (4.74)$$

that is applied for each n from 1 to N on the evolving image u. The numerical approximation scheme (4.74) is stable and consistent to the hyperbolic diffusion model (4.65), so it is also convergent and converges quite fast to its unique solution [38].

The proposed PDE-based inpainting class determines various hyperbolic interpolation models, depending on the selection of the parameters in (4.65)–(4.67) [38]. We have identified the best selection of these parameter values, which is

$$\alpha = 0.5, \lambda = 0.6, \beta = 1.7, \varepsilon = 1.6, \nu = 3,$$
$$k = 2, \zeta = 0.25, \gamma = 1.2, \eta = 1.3, \delta = 0.2, N = 29 \tag{4.75}$$

Our reconstruction approach inpaints successfully the missing part of the image, performing well in noisy conditions too. It reduces the additive noise and the unintended effects, and preserves the boundaries and other image features very well, given its hyperbolic character. Its processing speed and running time are influenced by the size of the inpainting domain, M.

Method comparison have been also performed. The described algorithm provides better structure-based interpolation results than some existing PDE inpainting approaches. See the averaged PSNR values obtained by our hyperbolic PDE-based technique and other reconstruction approaches, which are registered in Table 4.5 [38].

A structural inpainting example is provided in Fig. 4.11. The original image displayed in (a) is deteriorated by a missing region and additive Gausian noise in (b). It is reconstructed by the proposed hyperbolic PDE model in (c), by the Harmonic Inpainting model in (d), by TV Inpainting in (e), and by the inpainting scheme of Bertalmio et al. in (f).

4.3.3 Nonlinear Anisotropic Diffusion-Based Inpainting Framework

We have also developed an improved and more complex nonlinear anisotropic diffusion model for image inpainting. The second-order PDE-based interpolation framework disseminated in [40] consists of an anisotropic diffusion-based equation and some boundary conditions. The PDE model that recovers the original from the observed image is characterized by the next form:

$$\begin{cases} \frac{\partial u}{\partial t} - \psi^u(\|\nabla u\|)\nabla \cdot (\varphi_u(\|\nabla u\|)\nabla u) + \lambda(1 - 1_\Gamma)(u - u_0) = 0, & \forall(x, y) \in \Omega \\ u(x, y, 0) = u_0(x, y), & \forall(x, y) \in \Omega \\ u(t, x, y) = 0, & \forall(x, y) \in \partial\Omega \end{cases}$$

$$\tag{4.76}$$

Table 4.5 PSNR values achieved by several inpainting techniques

Interpolation method	Average PSNR (dB)
This hyperbolic PDE model	31.27
TV Inpainting	29.86
Harmonic Inpainting	28.42
Bertalmio et al. model	30.32

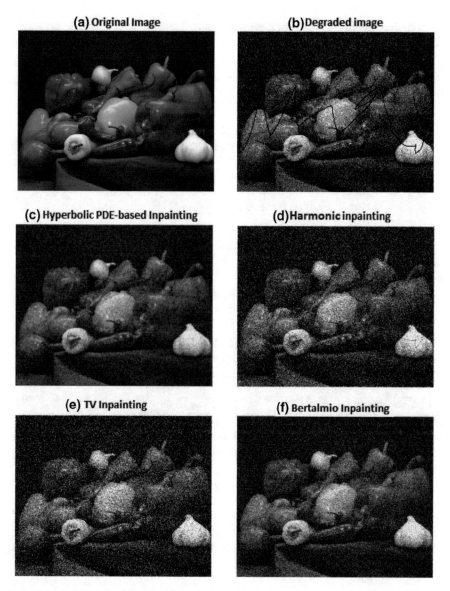

Fig. 4.11 Inpainting results achieved by our method and other schemes in noisy conditions

where $\lambda \in (0, 1]$ the image domain $\Omega \subseteq R^2$ and $\Gamma \subset \Omega$ represents the inpainting domain. This PDE-based model uses an *inpainting mask* which is provided by the characteristic function of the domain Γ: $1_\Gamma(x, y) = \begin{cases} 1, & if(x, y) \in \Gamma \\ 0, & if(x, y) \notin \Gamma \end{cases}$.

The first function of the anisotropic diffusion model (4.76) is modelled as following:

$$\begin{cases} \psi : (0, \infty) \to (0, \infty) \\ \psi''(s) = \gamma(\alpha s^r + \beta)^{\frac{1}{r+1}} \end{cases} \tag{4.77}$$

with $\alpha, \gamma \in (0, 3]$, $\beta \in (0, 3.5]$ and $r \in (0, 2]$

The diffusivity function proposed for this diffusion model has the following form:

$$\varphi_u : [0, \infty) \to [0, \infty)$$

$$\varphi_u(s) = \delta\left(\frac{\eta(u)}{\xi s^k + v \log 10(\eta(u))}\right)^{\frac{1}{k+1}} \tag{4.78}$$

where $\delta \in (0, 2)$ $\xi \in (1, 5]$ $v \in (0, 1)$ and $k \in \{1, 2, 3, 4\}$ Its conductance parameter is constructed as follows:

$$\eta(u(x, y, t)) = |\varepsilon\mu(\|\nabla u\|) + \zeta t| \tag{4.79}$$

where $\varepsilon > 1$ and $\varepsilon > 1$

The diffusivity function (4.78) is properly chosen, satisfying the main conditions required by a successful diffusion [40]. It is positive, since $\varphi_u(s) > 0, \forall s \in (0, \infty)$ This function is also monotonically decreasing, since $\varphi_u(s_1) = \delta\left(\frac{\eta(u)}{\xi s_1^k + v \log 10(\eta(u))}\right)^{\frac{1}{k+1}} \geq \varphi_u(s_2) = \delta\left(\frac{\eta(u)}{\xi s_2^k + v \log 10(\eta(u))}\right)^{\frac{1}{k+1}}$ for $\forall s_1 \geq s_2$. Also, it converges to zero when s goes to infinity: $\lim\limits_{s \to \infty} \varphi_u(s) = 0$ [40].

The component that uses the other function, $\psi''(\nabla u)$, has been introduced for controlling the speed of the diffusion process. It also has the role of better defining and preserving the image edges [40].

The reconstructed image is obtained by solving the nonlinear diffusion model (4.76). The well-posedness of this PDE-based scheme is carefully investigated in [40]. Thus, we have treated rigorously the mathematical validity of the considered anisotropic diffusion model, demonstrating the existence and uniqueness of a weak solution that would represent the optimal inpainting, under some certain assumptions [40].

Since we have:

$$\psi''(\|\nabla u\|^2)\nabla \cdot (\varphi_u(\|\nabla u\|^2)\nabla u) = div(\psi''(\|\nabla u\|^2)\varphi_u(\|\nabla u\|^2)\nabla u)$$
$$-2\varphi_u(\|\nabla u\|^2)\psi''(\|\nabla u\|^2)|\nabla u|^2\Delta u \tag{4.80}$$

and setting

$$g(s) = 2 \int_0^s \varphi_u(z) \psi^{u'}(z) z^2 dt, \quad \forall s > 0 \tag{4.81}$$

we get:

$$2\varphi_u(\|\nabla u\|^2)\psi^{u'}(\|\nabla u\|^2)|\nabla u|^2 \Delta u = div\big(g(\|\nabla u\|^2)\nabla u\big) - 2g'(\|\nabla u\|^2)|\nabla u|^2 \tag{4.82}$$

which leads to

$$\psi^u(\|\nabla u\|^2)div\big(\varphi_u(\|\nabla u\|^2)\nabla u\big) =$$
$$div\big((\psi^u(\|\nabla u\|^2)\varphi_u(\|\nabla u\|^2) - g(\|\nabla u\|^2))\nabla u\big) \tag{4.83}$$
$$-4\varphi_u(\|\nabla u\|^2)\psi^{u'}(\|\nabla u\|^2)\nabla u^2$$

Hence one can re-write the equation in (4.76) as:

$$\frac{\partial u}{\partial t} = div\Big(\big(\psi^u\big(\|\nabla u\|^2\big)\varphi_u\big(\|\nabla u\|^2\big) - g\big(\|\nabla u\|^2\big)\big)\nabla u\Big) + 4\varphi_u\big(\|\nabla u\|^2\big)\psi^{u'}\big(\|\nabla u\|^2\big)\|\nabla u\|^2 \tag{4.84}$$

We make the following assumption: the function $\psi^u(s^2)\varphi_u(s^2) - g(s^2)$ is convex and $\psi^u(s^2)\varphi_u(s^2) - g(s^2) \geq \rho > 0, \quad \forall s > 0$ ρ being a constant [40]. The function u represents a variational or weak solution to the Eq. (4.84) if it satisfies

$$\frac{\partial}{\partial t} \int_\Omega u(t, x, y)\chi(x, y)dxdy$$
$$+ \int_\Omega \big(\psi^u(\|\nabla u\|^2)\varphi_u(\|\nabla u\|^2) - g(\|\nabla u\|^2)\big)\nabla u \cdot \nabla\chi dxdy \tag{4.85}$$
$$-4 \int_\Omega \varphi_u(\|\nabla u\|^2)\psi^{u'}(\|\nabla u\|^2)\|\nabla u\|\chi dxdy = 0$$

for $(x, y) \in C^\infty(\Omega)$ $u \in L^2(0, T; H_0^1(\Omega))$ and $\frac{\partial u}{\partial t} \in L^2(0, T; H^{-1}(\Omega))$ where $H_0^1(\Omega)$ represents a Sobolev space while $H^{-1}(\Omega)$ represents its dual [40].

Let us set $v \in L^2(0, T; H_0^1(\Omega))$ and consider the following equation:

$$\begin{cases} \frac{\partial u}{\partial t} - div\big((\psi^u(\|\nabla u\|^2)\varphi_u(\|\nabla u\|^2) - g(\|\nabla u\|^2))\nabla u\big) = \\ 4\varphi_v(\|\nabla v\|^2)\psi^{v'}(\|\nabla v\|^2)\|\nabla v\|^2 \\ u(0, x, y) = u_0(x, y) \\ u(t, x, y) = 0 \ on \ (0, T) \times \partial\Omega \end{cases} \tag{4.86}$$

This problem given by (4.86) is parabolic in u and so it has a unique solution that satisfies $u \in L^2(0, T; H_0^1(\Omega))$ and $\frac{\partial u}{\partial t} \in L^2(0, T; H^{-1}(\Omega))$ Next one can prove that on a small interval the operator $v \rightarrow u$ represents a contraction in $L^2\big(0, T; H_0^1(\Omega)\big)$ [41].

One could demonstrate that (4.84), with $u_0 \in L^2(\Omega)$ admits an unique solution on some interval $(0, T)$ by applying the Banach's fixed point theorem [41]. Therefore, the proposed nonlinear anisotropic diffusion model is well-posed. Its solution, which represents the interpolated image, is then computed numerically by using a finite-difference method-based approximation algorithm [40].

The quantization of time and space coordinates provided by (2.13) is used here too [40]. The nonlinear partial differential equation from (4.76) is transformed into

$$\frac{\partial u}{\partial t} = \psi''(\|\nabla u\|)\left(\frac{\partial}{\partial x}(\varphi_u(\|\nabla u\|)u_x) + \frac{\partial}{\partial y}(\varphi_u(\|\nabla u\|)u_y)\right) - \lambda(1 - 1_\Gamma)(u - u_0)$$

(4.87)

and a robust finite difference-based discretization is next applied on Eq. (4.87) [40].

Thus, one determines $\psi_{i,j} = \psi''(u_{i,j})$ and $\varphi_{i,j} = \varphi_u(u_{i,j})$, where

$$u_{i,j} \approx \sqrt{\left(\frac{u_{i+h,j} - u_{i-h,j}}{2h}\right)^2 + \left(\frac{u_{i,j+h} - u_{i,j-h}}{2h}\right)^2}$$

(4.88)

represents the central difference-based discretization of the gradient magnitude [30, 31].

So, the term $\frac{\partial}{\partial x}(\varphi_u(\|\nabla u\|)u_x)$ is discretized spatially as $\varphi_{i+\frac{h}{2},j}(u_{i+h,j} - u_{i,j}) - \varphi_{i-\frac{h}{2},j}(u_{i,j} - u_{i-h,j})$ and $\frac{\partial}{\partial y}(\varphi_u(\|\nabla u\|)u_y)$ is approximated as $\varphi_{i,j+\frac{h}{2}}(u_{i,j+h} - u_{i,j}) - \varphi_{i,j-\frac{h}{2}}(u_{i,j} - u_{i,j-h})$ where:

$$\varphi_{i\pm\frac{h}{2},j} = \frac{\varphi_{i\pm h,j} + \varphi_{i,j}}{2}, \ \varphi_{i,j\pm\frac{h}{2}} = \frac{\varphi_{i,j\pm h} + \varphi_{i,j}}{2}$$

(4.89)

The next implicit discretization is obtained for (4.87), by using the forward difference for time derivative:

$$\frac{u_{i,j}^{n+\Delta t} - u_{i,j}^n}{\Delta t} = \psi_{i,j}\left(\varphi_{i+\frac{h}{2},j}(u_{i+h,j}^n - u_{i,j}^n) - \varphi_{i-\frac{h}{2},j}(u_{i,j}^n - u_{i-h,j}^n)\right.$$
$$\left. + \varphi_{i,j+\frac{h}{2}}(u_{i,j+h}^n - u_{i,j}^n) - \varphi_{i,j-\frac{h}{2}}(u_{i,j}^n - u_{i,j-h}^n)\right)$$
$$- \lambda(1 - 1_\Gamma)(u_{i,j}^n - u_{i,j}^0)$$

(4.90)

We can apply the parameter values $\Delta t = h = 1$ in (4.90). Therefore, it leads to the following explicit iterative numerical approximation scheme:

$$u_{i,j}^{n+1} = u_{i,j}^n\left(1 - \lambda(1 - 1_\Gamma) + \psi_{i,j}\left(\varphi_{i+\frac{1}{2},j} + \varphi_{i-\frac{1}{2},j} + \varphi_{i,j+\frac{1}{2}} + \varphi_{i,j-\frac{1}{2}}\right)\right)$$
$$+ u_{i+1,j}^n\psi_{i,j}\varphi_{i+\frac{1}{2},j} + u_{i-1,j}^n\psi_{i,j}\varphi_{i-\frac{1}{2},j} + u_{i,j+1}^n\psi_{i,j}\varphi_{i,j+\frac{1}{2}}$$
$$+ u_{i,j-1}^n\psi_{i,j}\varphi_{i,j-\frac{1}{2}} + u_{i,j}^0\lambda(1 - 1_\Gamma)$$

(4.91)

where $u^n_{0,j} = u^n_{1,j}, u^n_{I,j} = u^n_{I+1,j}, u^n_{i,0} = u^n_{i,1}, u^n_{i,J} = u^n_{i,J+1}$, for $n = 0,1, ...,N$.

The iterative approximation algorithm (4.91) is consistent to the PDE-based model (4.76) and converges quite fast to its solution representing the inpainted image. It has been successfully used in the interpolation experiments performed by us [40].

Thus, we have tested the proposed inpainting technique on hundreds images corrupted by noise and missing regions, getting good results. Some of the image datasets used in our experiments are the three volumes of the USC–SIPI database [40].

The performance of our reconstruction approach has been assessed by applying similarity metrics such as PSNR (Peak Signal to Noise Ratio), SNR (Signal to Noise Ratio) and MSE (Mean-Squared Error) [42]. The next set of values for the PDE model's coefficients, which lead to an optimal image inpainting, have been detected empirically, using the trial and error method with these performance measures:

$$\lambda = 0.7, \gamma = 0.6, \alpha = 1.2, \beta = 1.7, r = 0.45, \delta = 0.5, \xi = 3.5,$$
$$\nu = 0.4, k = 2, \varepsilon = 1.7, \zeta = 0.3 \tag{4.92}$$

Since the numerical approximation scheme converges fast to the optimal inpainting, its number of iterations, N, being quite low, the execution time is also low. However, it increases proportionally with the size of the inpainting area and the amount of image noise [40].

The proposed nonlinear diffusion-based approach inpaints successfully the damaged images by directing (and also controlling) the speedy diffusion process to inpainting domain while preserving their features (edges, corners and other details). It performs successfully in both normal and noisy image conditions. It reduces the amount of additive Gaussian noise and the undesired image effects.

One can see some interpolation results obtained by the described technique in Fig. 4.12. The *Lenna* image affected by a missing part (b) is inpainted successfully in normal conditions in c) after $N = 29$ iterations, while the successful filling of the same missing zone (black scratch) in conditions of additive noise characterized by $\mu = 0.21$ and *variance* = 0.02 requires $N = 39$ time steps.

The performed method comparison also proves the effectiveness of the developed method. This anisotropic diffusion-based inpainting framework has been compared to some well-known PDE and variational image reconstruction models [40].

So, our approach outperforms second-order PDE-based methods, such as Harmonic Inpainting, TV Inpainting and the interpolation models derived from nonlinear diffusion-based schemes like the Perona-Malik algorithm, achieving lower MSE and higher PSNR values. It also obtains comparable good results to some fourth-order PDE inpainting schemes, such as the $TV - H^{-1}$ Inpainting model described in 4.1.2.

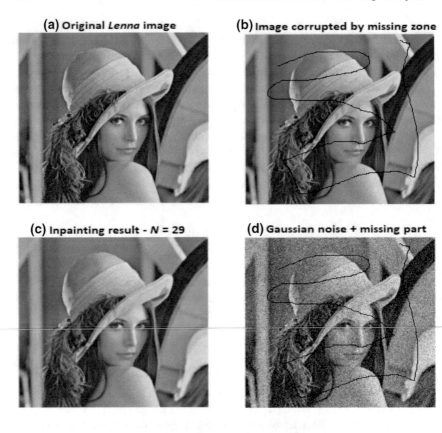

(a) Original *Lenna* image

(b) Image corrupted by missing zone

(c) Inpainting result - *N* = 29

(d) Gaussian noise + missing part

(e) Inpainting in noisy conditions: *N* = 39

Fig. 4.12 Inpainting results achieved by the proposed model in normal and noisy conditions

Table 4.6 Method comparison: average PSNR and MSE values

Inpainting algorithm	Average PSNR (dB)	Average MSE
The proposed AD Inpainting	36.1338	15.8380
Harmonic Inpainting	29.6887	69.8576
Total Variation Inpainting	34.2561	24.4047
TV $-$ H^{-1} Inpainting	36.3805	14.9634

Some method comparison results are presented in the following table and figure. The average PSNR and MSE values achieved by several techniques are registered in Table 4.6. The TV $-$ H^{-1} Inpainting scheme produces slightly better average values of these performance measures, the AD (anisotropic diffusion)—based scheme proposed by us is converging faster than it and has a considerably lower running time [40].

An inpainting method comparison example involving the approaches mentioned in the above table is also described in Fig. 4.13. The figure depicts the interpolation results achieved by these techniques on the [512 × 512] *Barbara* image that is deteriorated by two missing regions representing black hand-written texts.

The proposed reconstruction framework requires much fewer iterations than the other methods to reach the optimal interpolation. Our algorithm inpaints successfully the image after 45 steps, while TV- Inpainting requires 850 iterations and TV $-$ H^{-1} Inpainting needs 1500 iterations.

While the proposed nonlinear anisotropic diffusion-based inpainting technique provides an effective structural image reconstruction, it has also its own shortcomings. Thus, it performs considerably weaker the texture-based image inpainting tasks.

Therefore, as part of our future research in this field, we intend to further improve this image reconstruction framework, so that to become able to inpaint the damaged textures, too.

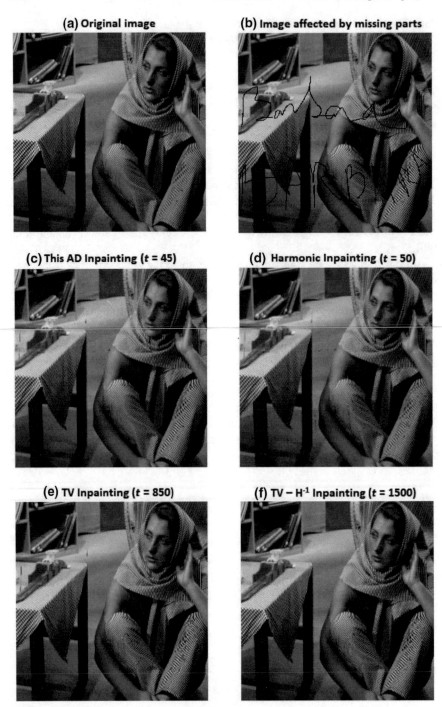

Fig. 4.13 Interpolation output produced by several PDE-based algorithms

References

1. C.B. Schonlieb, in *Partial Differential Equation Methods for Image Inpainting*, vol. 29. (Cambridge University Press, 2015)
2. A.A. Efros, T.K. Leung, Texture synthesis by non-parametric sampling. Proc. Int. Conf. Comput. Vis. **2**, 1033–1038 (1999)
3. H. Igehy, L. Pereira, Image replacement through texture synthesis. Proc. Int. Conf. Image Process. **3**, 186–189 (1997)
4. A. Criminisi, P. Perez, K. Toyama, Region filling and object removal by exemplar-based image inpainting. IEEE Trans. Image Process. **13**(9), 1200–1212 (2004)
5. V. Casselles, Exemplar-based image inpainting and application. SIAM News 44(10) (2011)
6. B. Song, *Topics in Variational PDE Image Segmentation, Inpainting and Denoising* (University of California, 2003)
7. S. Esedoglu, J. Shen, Digital inpainting based on the Mumford-Shah Euler image model. Eur. J. Appl. Math. (2002)
8. L. Ambrosio, V.M. Tortorelli, Approximation of functionals depending on jumps by elliptic functionals via Γ—convergence. Comm. Pure Appl. Math. **43**, 999–1036 (1990)
9. T.F. Chan, J. Shen, *Morphologically Invariant PDE Inpaintings*, UCLA CAM Report (2001), pp. 1–15
10. P. Getreuer, Rudin–Osher–Fatemi total variation denoising using Split Bregman. Image Process. Line (2012)
11. P. Getreuer, Total variation inpainting using Split Bregman. Image Process. Line **2**, 147–157 (2012)
12. K. Papafitsoros, C.B. Schoenlieb, B. Sengul, Combined first and second order total variation inpainting using split Bregman. Image Process. On Line **2013**, 112–136 (2013)
13. K. Papafitsoros, C.B. Schonlieb, A combined first and second order variational approach for image reconstruction. J. Math. Imaging Vision **48**, 308–338 (2014)
14. K. Bredies, K. Kunisch, T. Pock, Total generalized variation. SIAM J. Imaging Sci. **3**, 1–42 (2009)
15. K. Bredies, M. Holler, A TGV-based framework for variational image decompression, zooming, and reconstruction. Part I: analytics. SIAM J. Imaging Sci. **8**(4), 2814–2850 (2015)
16. K. Bredies, M. Holler, A TGV-based framework for variational image decompression, zooming, and reconstruction. Part II: numerics. SIAM J. Imaging Sci **8**(4), 2851–2886 (2015)
17. M.V. Afonso, J.M.R. Sanches, Blind inpainting using l_0 and total variation regularization. IEEE Trans. Image Process. **24**(7), 2239–2253 (2015)
18. M. Neri, E.R. Zara, Total variation-based image inpainting and denoising using a primal-dual active set method. Philippine Sci. Lett. **7**(1), 97–103 (2014)
19. T.F. Chan, S.-H. Kang, J. Shen, Euler's elastica and curvature based inpaintings. SIAM J. Appl. Math. (2002)
20. S. Masnou, J.-M. Morel, Level-lines based disocclusion, in *Proceedings of 5th IEEE International Conference on Image Processing*, vol. 3 (Chicago, 1998), pp. 259–263
21. M. Bertalmio, G. Sapiro, V. Caselles, C. Ballesters, Image inpainting, in *Proceedings of ACM Conference Computer Graphics* (*SIGGRAPH,* New Orleans, LU, July 2000), pp. 417–424
22. M. Bertalmio, A. L. Bertozzi, G. Sapiro, Navier-stokes, fiuid dynamics, and image and video inpainting, in *Proceedings of Conference Computer Vision Pattern Recognition* (Hawai, December 2001), pp. 355–362
23. M.A. Ebrahimi, M. Holst, E. Lunasin, The Navier-Stokes-Voight model for image inpainting. IMA J. Appl. Math. **78**(5), 869–894 (2013)
24. T.F. Chan, J. Shen, Non-texture inpainting by curvature-driven diffusions (CDD). J. Visual Comm. Image Rep. **4**(12), 436–449 (2001)
25. M. Burger, L. He, C. Schonlieb, Cahn-Hilliard inpainting and a generalization for grayvalue images. SIAM J. Imaging Sci. **2**(4), 1129–1167 (2009)

26. S. Osher, A. Sole, L. Vese, Image decomposition and restoration using total variation minimization and the H^{-1} norm. Multiscale Model. Simul. SIAM Interdiscip. J. **1**(3), 349–370 (2003)
27. C.B. Schoenlieb, A. Bertozzi, Unconditionally stable schemes for higher order inpainting. Comm. Math. Sci. **2**(9), 413–457 (2011)
28. T. Barbu, Variational image inpainting technique based on nonlinear second-order diffusions. Comput. Electr. Eng. **54**, 345–353 (2016)
29. G. Arfken, The method of steepest descents, §7.4, in *Mathematical Methods for Physicists, 3rd ed.* (Orlando, FL: Academic Press, 1985), pp. 428–436
30. D. Gleich, *Finite Calculus: A Tutorial for Solving Nasty Sums* (Stanford University, 2005)
31. P. Johnson, *Finite Difference for PDEs* (School of Mathematics, University of Manchester, Semester I, 2008)
32. A. Kalaitzis, *Image Inpainting with Gaussian Processes*, Master of Science Thesis, School of Informatics, University of Edinburgh, 2009
33. F.A. Jassim, Image inpainting by kriging interpolation technique. World Comput. Sci. Info. Technol. J. **3**(5), 91–96 (2013)
34. T. Barbu, A novel variational framework for structural image completion, in *Proceedings of Joint International Conference OPTIM-ACEMP 2017* (Brasov, Romania, May 2017, IEEE), pp. 815–820, 25–27
35. T. Barbu, Hybrid image interpolation technique based on nonlinear second and fourth-order diffusions, in *Proceedings of the 13th International Symposium on Signals, Circuits and Systems, ISSCS'17* (Iasi, Romania, July 13–14, 2017), pp. 1–5
36. T. Barbu, I. Munteanu, A well-posed second-order anisotropic diffusion-based structural inpainting scheme. ROMAI J. ROMAI Soc. **1** (2017)
37. T. Barbu, Nonlinear anisotropic diffusion-based structural inpainting framework, in *Proceedings of the 13th International Conference on Advanced Technologies, Systems and Services in Telecommunications, TELSIKS '17* (Nis, Serbia, 18–20 October 2017, IEEE), pp. 207–210
38. T. Barbu, Structural image interpolation using a nonlinear second-order hyperbolic PDE-based model, in *Proceedings of the 6th IEEE International Conference on e-Health and Bioengineering, EHB 2017* (Sinaia, Romania, 22–24 June 2017), pp. 5–8
39. A.P. Witkin, Scale-space filtering, in *Proceedings of the Eighth International Joint Conference on Artificial Intelligence*, vol. 2 (IJCAI '83, Karlsruhe, 8–12 August 1983), pp. 1019–1022
40. T. Barbu, Second-order Anisotropic Diffusion-based Framework for Structural Inpainting, in *Proceedings of the Romanian Academy, Series A: Mathematics, Physics, Technical Sciences, Information Science*, vol. 19, Issue 2 (April–June 2018)
41. V. Barbu, *Nonlinear Differential Equations of Monotone Type in Banach Spaces* (Springer, Berlin, 2010)
42. E. Silva, K.A. Panetta, S.S. Agaian, Quantify similarity with measurement of enhancement by entropy, in *Proceedings: Mobile Multimedia/Image Processing for Security Applications, SPIE Security Symposium 2007*, vol. 6579 (April 2007), pp. 3–14

Chapter 5
Conclusions

The topic of this book is located at the intersection of partial differential equations, image processing and analysis, and numerical analysis. This work has brought many theoretical and practical results in two important and closely related image processing domains, namely the diffusion-based image denoising and inpainting, which have been addressed here.

Both the state of the art techniques of these fields and the novel approaches developed by us in these areas have been surveyed and compared in this manuscript. Thus, we have performed a high amount of research in the PDE-based image processing and computer vision fields in the last 15 years. The results of this research have been disseminated in numerous articles published in some recognized international journals and volumes of international scientific events. Many of these publications are listed in the references' section and widely cited throughout this book. They disseminate our most important and recent PDE-based image restoration and inpainting solutions.

As already mentioned, the nonlinear partial differential equations represent a powerful image denoising and reconstruction tool, since they solve properly a major challenge in these domains, which is the overcoming of the undesirable blurring effect and the preservation of the essential features. Nonlinear PDE-based models have been considered for both restoration and inpainting tasks in this book, but linear PDE-based schemes have been also applied in the restoration case.

Although the existing linear PDE-based models have some obvious drawbacks, such as blurring and dislocating the edges when moving from finer to coarser scales, we have included a chapter devoted to the linear diffusion-based denoising in this manuscript. That is because we have developed some improved linear PDE-based image denoising techniques that alleviate the unintended effects, have the localization property and outperform not only the simple linear diffusion models like that provided by the heat equation, but also some nonlinear second-order PDE algorithms. Such an improved linear PDE restoration scheme has been derived from a SDE-based image denoising framework developed and investigated by us.

© Springer International Publishing AG, part of Springer Nature 2019
T. Barbu, *Novel Diffusion-Based Models for Image Restoration and Interpolation*,
Signals and Communication Technology,
https://doi.org/10.1007/978-3-319-93006-0_5

The three book chapters describing the two PDE-based image processing fields have provided comprehensive overviews of these domains. So, the most state-of-the-art linear and nonlinear diffusion-based restoration techniques, and nonlinear PDE-based interpolation approaches, using variational or non-variational differential models, have been surveyed in these chapters. Thus, some influential linear PDE schemes and nonlinear second and fourth order diffusion-based models for image restoration, as well as state of the art variational second-order PDE models and non-variational high-order PDE-based schemes for image interpolation have been described. Some existing hybrid diffusion-based denoising and inpainting solutions have also been presented.

Since our recent research contributions in these fields represent the main focus of this book, these surveys have been followed by the sections describing the theoretical and practical research results achieved by us, in each of the three chapters. Our novel image restoration and inpainting approaches have been grouped by categories and sub-categories in each of those chapter. Thus, since each technique considered here is based on a mathematical model representing a partial differential equation with some boundary conditions or an energy functional minimization scheme that leads to a PDE model, our approaches have been divided on the basis of several criteria, such as the linearity (linear or nonlinear), the variational or non-variational character, the order of the corresponding differential equation (second or fourth order), the character of that equation (parabolic or hyperbolic) and the individual or hybrid character of that image restoration or interpolation method.

Each diffusion-based denoising or inpainting model described in this book has been mathematically investigated, numerical solved and implemented. Various finite difference-based numerical approximation algorithms constructed for these models have been presented here and numerical experiments and method comparison have been discussed for each diffusion scheme, their results being displayed in the many tables and figures provided in this book.

Obviously, not all of our techniques disseminated in the papers listed in the references section and cited throughout the book have been equally treated in this manuscript. So, full descriptions have been provided for the most important of them in terms of the performance and originality [1–9], while some works have been barely mentioned here [10–12].

Other restoration and inpainting methods have been described more briefly in this manuscript, by omitting or not detailing in full some parts of them, such as some mathematical computations and investigations, or the numerical approximations. Thus, some long mathematical treatments, such as those related to the PDE model's well-posedness investigations, have been shortened and the reader has been referred to other sources [6, 13–17]. In other cases, the numerical approximation processes of some diffusion-based schemes have been briefly presented or even entirely omitted, because of their similarity to other numerical discretization procedures, already described in the book [16, 18–20]. Some mathematical procedures for deriving PDE-based models from the variational algorithms have been also briefed.

As already mentioned in this book, our denoising and interpolation techniques described here have some limits with respect to the tasks they can perform. Thus, our diffusion-based restoration models remove successfully the additive Gaussian noise, but are not appropriate for other types of image noise. However, they can be properly adapted to other noises, such as Poisson noise and Laplacian noise. So, we intend to focus on developing such diffusion-based filtering solutions for other types of noise in the future.

Also, our nonlinear PDE-based interpolation techniques described in the fourth chapter provide a successful structural reconstruction, but cannot inpaint properly the image textures. Modelling some effective texture-based inpainting solutions represents another task of our future research.

The presented reconstruction approaches can be further improved by automatizing them. So, their inpainting masks would be automatically detected, by applying some object detection techniques, instead of constructed manually. We developed such segmentation and object detection solutions, some of them using PDE models [21], in the past, which could be applied for this task.

Also, the edge detection, the image segmentation and the image and video object detection domains represent obvious application areas of our nonlinear second-order diffusion-based algorithms. Since these techniques have a strong edge-preserving and edge-enhancing character, some new effective edge detection solutions can be obtained using them [22]. These edge detection models could then lead to image segmentation and object detection solutions. And a closely related domain that is the image and video object removal represents an important application field of our diffusion-based inpainting methods.

References

1. T. Barbu, Linear Hyperbolic Diffusion-based Image Denoising Technique, in *Lecture Notes in Computer Science Proceeding of the 22nd International Conference on Neural Information Processing, ICONIP 2015*, Part III, Istanbul, Turkey, ed. by S. Arik et al. vol. 9491, (Springer, November 9–12, 2015), pp. 471–478
2. T. Barbu, A linear diffusion-based image restoration approach. *ROMAI J.* (2) (2015), pp. 133–139 (ROMAI Society)
3. T. Barbu, A. Favini, Novel stochastic differential model for image restoration, in *Proceedings of the Romanian Academy, Series A: Mathematics, Physics, Technical Sciences, Information Science*, vol. 17, num. 2, pp. 109–116, April-June 2016
4. T. Barbu, Nonlinear PDE model for image restoration using second-order hyperbolic equations. *Numer. Funct. Analysis and Optimization*, **36**(11), 1375–1387 (November 2015) (Taylor & Francis)
5. T. Barbu, A nonlinear second-order hyperbolic diffusion scheme for image restoration, *U.P.B. Sci. Bull. Ser. C* **78**(2) (2016)
6. T. Barbu, PDE-based restoration model using nonlinear second and fourth order diffusions, in *Proceedings of the Romanian Academy, Series A: Mathematics, Physics, Technical Sciences, Information Science*, vol. 16, num. 2 (April–June 2015), pp. 138–146
7. T. Barbu, A Hybrid Nonlinear Fourth-order PDE-based Image Restoration Approach, in *Proceedings of 20th International Conference on System Theory, Control and Computing ICSTCC 2016*, Sinaia, Romania, October 13–15 (IEEE, 2016), pp. 761–765
8. T. Barbu, Variational image inpainting technique based on nonlinear second-order diffusions. Comput. Electr. Eng. **54**, 345–353 (2016)

9. T. Barbu, Second-order anisotropic diffusion-based framework for structural inpainting, in *Proceedings of the Romanian Academy, Series A: Mathematics, Physics, Technical Sciences, Information Science*, vol. 19, Issue 2 (April–June 2018)

10. T. Barbu, Novel linear image denoising approach based on a modified Gaussian filter kernel. *Numer. Funct. Anal. Optim.* **33**(11), 1269–1279 (2012) ((Taylor & Francis Group, LLC)

11. T. Barbu, A novel variational PDE technique for image denoising, in *Proceeding of the 20th International Conference on Neural Information Processing, ICONIP 2013*, part III, Daegu, Korea, 3–7 November, 2013, vol. 8228, ed. by M. Lee et al. Lecture Notes in Computer Science (Springer-Verlag Berlin Heidelberg, 2013) pp. 501–508

12. T. Barbu, PDE-based image restoration using variational denoising and inpainting models, in *Proceedings of the 18th International Conference on System Theory, Control and Computing, ICSTCC 2014*, Sinaia, Romania, October 17–19 (2014), pp. 694–697

13. T. Barbu, A. Favini, Rigorous mathematical investigation of a nonlinear anisotropic diffusion-based image restoration model. Electron. J. Differ. Equ. **2014**(129), 1–9 (2014)

14. T. Barbu, C. Morosanu, *Image Restoration using a Nonlinear Second-order Parabolic PDE-based Scheme*, vol. XXV, Fasc. 1 (Analele Stiintifice ale Universitatii Ovidius Constanta, Seria Matematică, 2017) pp. 33–48

15. T. Barbu, I. Munteanu, A nonlinear fourth-order diffusion-based model for image denoising and Restoration, in *Proceedings of the Romanian Academy, Series A: Mathematics, Physics, Technical Sciences, Information Science*, vol. 18, Issue 2 (April–June 2017) pp. 108–115

16. T. Barbu, G. Marinoschi, Image denoising by a nonlinear control technique. Int. J. Control **90**(5), 1005–1017 (2017). (Taylor & Francis)

17. T. Barbu and I. Munteanu, A well-posed second-order anisotropic diffusion-based structural inpainting scheme, *ROMAI Journal*, ROMAI Society, No. 1, 2017

18. T. Barbu, Nonlinear fourth-order diffusion-based model for image denoising, in *Soft Computing Applications. Advances in Intelligent Systems and Computing*, vol. 633, ed. by V. Balas, L. Jain, M. Balas (Springer, Cham, September 2017), pp. 423–429

19. T. Barbu, Nonlinear fourth-order diffusion-based image restoration scheme. ROMAI J. **1**, 1–8 (2016). (ROMAI Society)

20. T. Barbu, Nonlinear anisotropic diffusion-based structural inpainting framework, in *Proceedings of the 13th International Conference on Advanced Technologies, Systems and Services in Telecommunications, TELSIKS '17*, Nis, Serbia (IEEE, 18–20 October 2017), pp. 207–210

21. T. Barbu, Robust contour tracking model using a variational level-set algorithm. Numerical Functional Analysis and Optimization **35**(3), 263–274 (2014). (Pub. Taylor & Francis)

22. T. Barbu, A nonlinear anisotropic diffusion-based edge detection scheme, in *Proceedings of the 17th International Multidisciplinary Scientific GeoConference, SGEM 2017*, Albena, Bulgaria (27 June–6 July 2017), pp. 21–28

Printed in the United States
By Bookmasters